The Excel Manual

Beverly Dretzke

STATISTICS
Informed Decisions Using Data

Michael Sullivan III

Upper Saddle River, NJ 07458

Editor-in-Chief: Sally Yagan
Supplement Editor: Joanne Wendelken
Assistant Managing Editor: John Matthews
Production Editor: Wendy A. Perez
Supplement Cover Manager: Paul Gourhan
Supplement Cover Designer: Joanne Alexandris
Manufacturing Buyer: Ilene Kahn

Pearson Education, Inc.
Upper Saddle River, NJ 07458

Printed in the United States of America

10 9 8 7 6 5 4 3 2 1

ISBN 0-13-046495-3

Pearson Education Ltd., *London*
Pearson Education Australia Pty. Ltd., *Sydney*
Pearson Education Singapore, Pte. Ltd.
Pearson Education North Asia Ltd., *Hong Kong*
Pearson Education Canada, Inc., *Toronto*
Pearson Educación de Mexico, S.A. de C.V.
Pearson Education—Japan, *Tokyo*
Pearson Education Malaysia, Pte. Ltd.
Pearson Education, *Upper Saddle River, New Jersey*

Contents:

Getting Started with Microsoft Excel

Overview

This manual is intended as a companion to Sullivan's *Statistics: Informed Decisions Using Data*. The manual presents instructions on how to use Microsoft Excel to carry out selected examples and exercises from *Statistics: Informed Decisions Using Data*.

The first section of the manual contains an introduction to Microsoft Excel and how to perform basic operations such as entering data, using formulas, saving worksheets, retrieving worksheets, and printing. All the screens pictured in this manual were obtained using the Office 2000 version of Microsoft Excel on a PC. You may notice slight differences if you are using a different version or a different computer.

Getting Started with the User Interface

GS 1.1 The Mouse

The mouse is a pointer device that allows you to move around the Excel worksheet and to select specific locations and objects. There are four main mouse operations: Select, click, double-click, and right-click.

1. To **select** generally means to move the mouse pointer so that the white arrow is pointing at or is positioned directly over an object. You will often **select** commands in the standard toolbar located near the top of the screen. Some of the more familiar of these commands are open, save, and print.

2. To **click** means to press down on the left button of the mouse. You will frequently select cells of the worksheet and commands by "clicking" the left button.

3. To **double-click** means to press the left mouse button twice in rapid succession.

4. To **right-click** means to press down on the right button of the mouse. A right-click is often used to display special shortcut menus.

GS 1.2 The Excel Worksheet

The figure shown below presents the Office 2000 version of a blank Excel worksheet. Important parts of the worksheet are labeled.

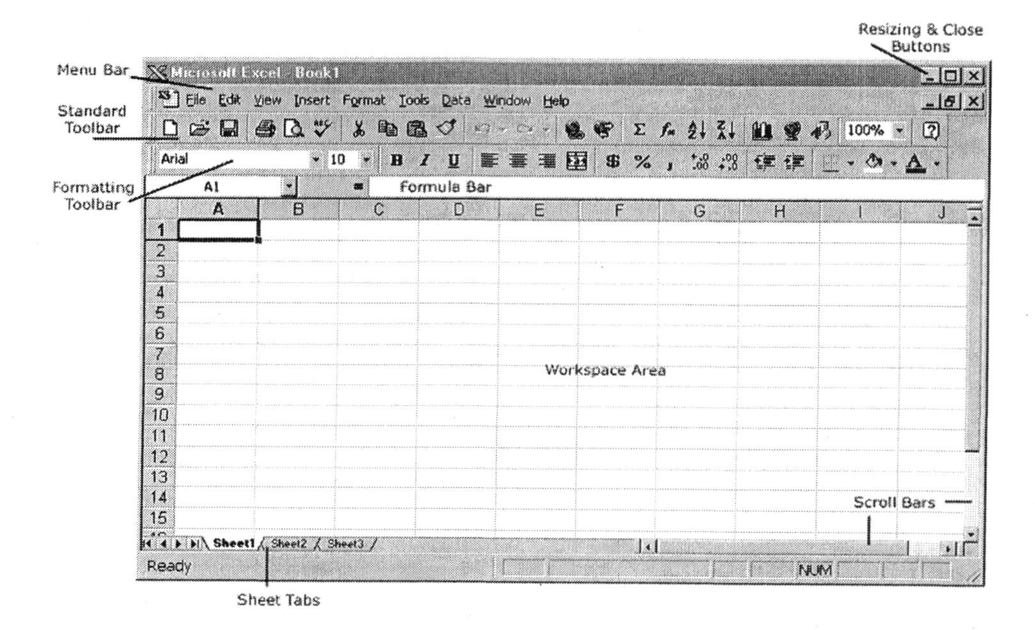

GS 1.3 Menu Conventions

Excel uses standard conventions for all menus. For example, the Menu bar contains the commands File, Edit, View, etc. Selecting one of these commands will "drop down" a menu. The Edit menu is displayed at the top of the next page.

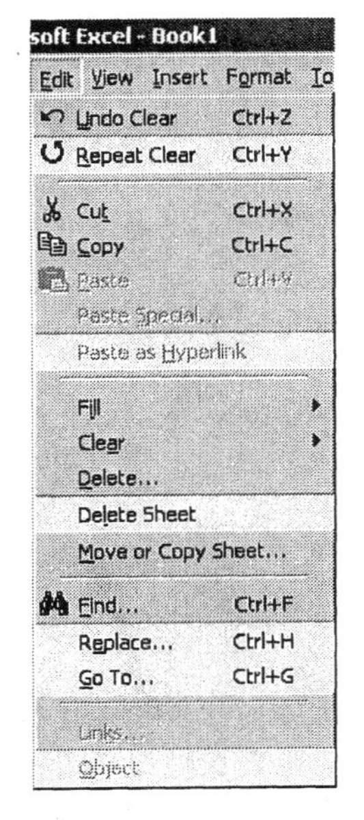

Icons to the left of the Cut, Copy, Paste, and Find commands indicate toolbar buttons that are equivalent to the menu choices.

Keyboard shortcuts are displayed to the right of the commands. For example, Ctrl+X is a keyboard shortcut for Cut.

The triangular markers to the right of Fill and Clear indicate that selection of these commands will result in a second menu of choices.

Selection of commands that are followed by an ellipsis (e.g., Paste Special… and Delete…) will result in the display of a dialog box that usually must be responded to in some way in order for the command to be executed.

The menus found in other locations of the Excel worksheet will operate in the same way.

GS 1.4 Dialog Boxes

Many of the statistical analysis procedures that are presented in this manual are associated with commands that are followed by dialog boxes. Dialog boxes usually require that you select from alternatives that are presented or that you enter your choices.

For example, if you click **Insert** and select **Function**, a dialog box like the one shown below will appear. You are required to select both a Function category and a Function name. You make your selections by clicking on them.

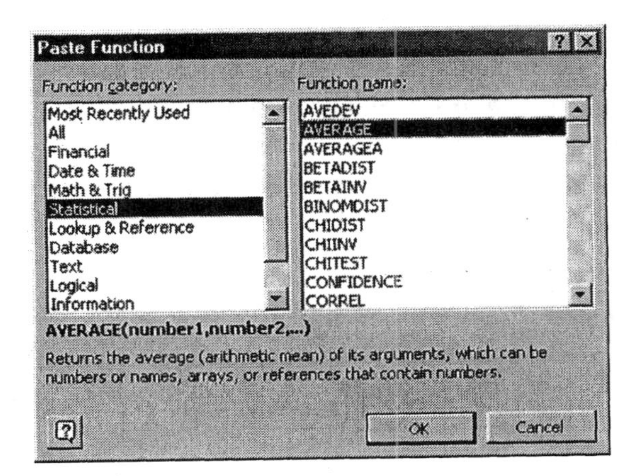

When you click the OK button at the bottom of the dialog box, another dialog box will often be displayed that asks you to provide information regarding location of the data in the Excel worksheet

Getting Started with Opening Files

GS 2.1 Opening a New Workbook

When you start Excel, the screen will open to **Sheet 1** of **Book 1**. Sheet names appear on tabs at the bottom of the screen. The name "Book 1" will appear in the top left corner.

If you are already working in Excel and have finished the analyses for one problem and would like to open a new book for another problem, follow these steps: First, at the top of the screen, click **File** and select **New**. Next, click **OK** in the New dialog box. If you were previously working in Book1, the new worksheet will be given the default name Book 2.

The names of books opened during an Excel work session will be displayed at the bottom of the Window menu. To return to one of these books, click **Window** and then click the book name.

GS 2.2 Opening a File That Has Already Been Created

To open a file that you or someone else has already created, click on **File** and select **Open**. A list of file locations will appear. Select the location by clicking on it. Many of the data files that are presented in your statistics textbook are available a CD-ROM that accompanies this manual. To open any of these files, you will select the appropriate CD drive on your computer. This drive will be designed with a symbol that looks like .

After you select the CD drive, a list of folders and files available on the CD will appear. You will need to select the folder or file you want by clicking on it. If you have selected a folder, another screen will appear with a list of files contained in the folder. Click on the name of the file that you would like to open.

Getting Started with Entering Information

GS 3.1 Cell Addresses

Columns of the worksheet are identified by letters of the alphabet and rows are identified by numbers. The cell address A1 refers to the cell located in column A row 1. The dark outline around a cell means that it is "active" and is ready to receive information. In the figure shown below, cell C1 is ready to receive information. You can also see C1 in the **Name Box** to the left of the **Formula Bar**. You can move to different cells of the worksheet by using the mouse pointer and clicking on a cell. You can also press [**Tab**] to move to the right or left, or you can use the arrow keys on the keyboard.

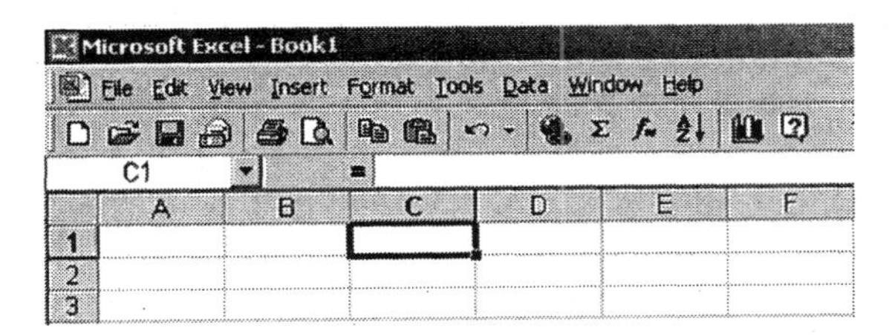

You can also activate a **range** of cells. To activate a range of cells, first click in the top cell and drag down and across (or click in the bottom cell and drag up and across). The range of cells highlighted in the figure below is designated B2:D6.

GS 3.2 Types of Information

Three types of information may be entered into an Excel worksheet.

1. **Text**. The term "text" refers to alphabetic characters or a combination of alphabetic characters and numbers, sometimes called "alphanumeric." The figure provides an

example of an entry comprised solely of alphabetic characters (cell A1) and an entry comprised of a combination of alphabetic characters and numbers (cell B1).

	A	B	C	D	E	F	G
1	Sue Clark	25 years					
2							
3							

2. **Numeric**. Any cell entry comprised completely of numbers falls into the "numeric" category.

3. **Formulas**. Formulas are a convenient way to perform mathematical operations on numbers already entered into the worksheet. Specific instructions are provided in this manual for problems that require the use of formulas.

GS 3.3 Entering Information

To enter information into a cell of the worksheet, first activate the cell. Then key in the desired information and press [**Enter**]. Pressing the [Enter] key moves you down to the next cell in that column. The information shown below was entered as follows:

1. Click in cell **A1**. Key in **1**. Press [**Enter**].

2. Key in **2**. Press [**Enter**].

3. Key in **3**. Press [**Enter**].

	A	B	C	D	E	F	G
1	1						
2	2						
3	3						
4							
5							

GS 3.4 Using Formulas

When you want to enter a formula, begin the cell entry with an equal sign (=). The arithmetic operators are displayed at the top of the next page.

Arithmetic operator	Meaning	Example
+	Addition	3+2
-	Subtraction	3-2
*	Multiplication	3*2
/	Division	3/2
^	Exponentiation	3^2

Numbers, cell addresses, and functions can be used in formulas. For example, to sum the contents of cells A1 and B1, you can use the formula =A1+B1. To divide this sum by 2, you can use the formula =(A1+B1)/2. Note that Excel carries out expressions in parentheses first and then uses the results to complete the calculations of the formula. Formulas will sometimes not produce the desired results because parentheses were necessary but were not used.

Getting Started with Changing Information

GS 4.1 Editing Information in the Cells

There are several ways that you can edit information that has already been entered into a cell.

- If you have not completed the entry, you can simply backspace and start over. Clicking on the red X to the left of the Formula Bar will also delete an incomplete cell entry.

- If you have already completed the entry and another cell is activated, you can click on the cell you want to edit and then press either [**Delete**] or [**Backspace**] to clear the contents of the cell.

- If you want to edit part of the information in a cell instead of deleting all of it, follow the instructions provided in the example.

1. Let's say that you wanted to enter 1234 in cell A1 but instead entered 124. Return to cell **A1** to make it the active cell by either clicking on it with the mouse or by using the arrow keys.

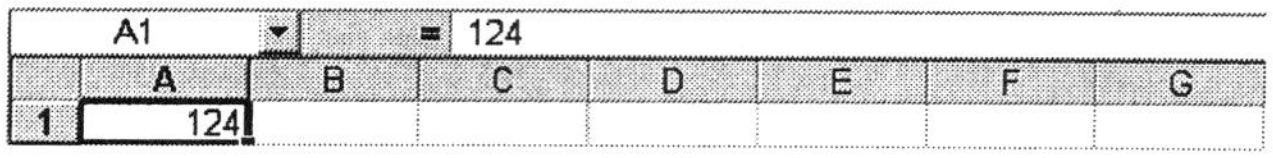

A1	= 124						
	A	B	C	D	E	F	G
1	124						

2. You will see A1 in the Name Box and 124 in the Formula Bar. Click between 2 and 4 in the Formula Bar so that the **I-beam** is positioned there.

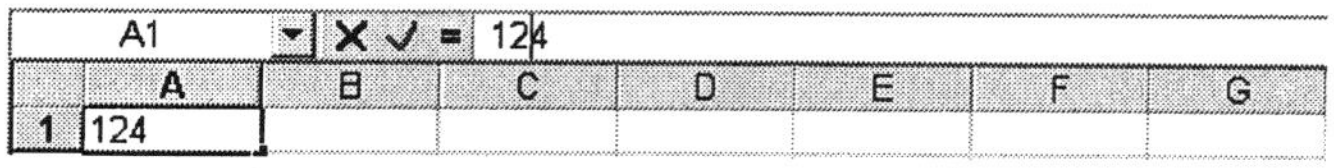

A1	X ✓ = 124						
	A	B	C	D	E	F	G
1	124						

3. Enter the number 3 and press [**Enter**].

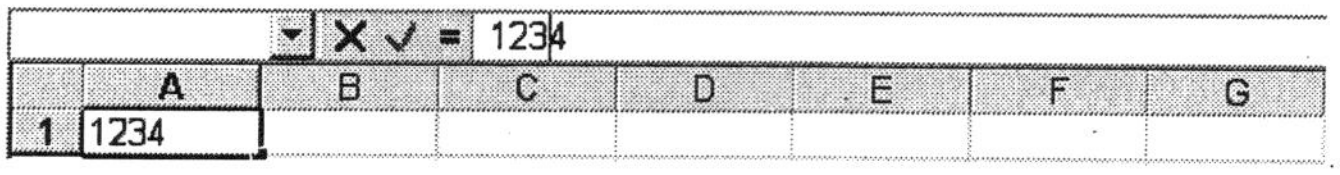

	X ✓ = 1234						
	A	B	C	D	E	F	G
1	1234						

GS 4.2 Copying Information

To copy the information in one cell to another cell, follow these steps:

1. First click on the source cell. Then, at the top of the screen, click **Edit** and select **Copy**.

2. Click on the target cell where you want the information to be placed. Then, at the top of the screen, click **Edit** and select **Paste**.

To copy a range of cells to another location in the worksheet, follow these steps:

1. First click and drag over the range of cells that you want to copy so that they are highlighted. Then, at the top of the screen, click **Edit** and select **Copy**.

2. Click in the topmost cell of the target location. Then, at the top of the screen, click **Edit** and select **Paste**.

To copy the contents of one cell to a range of cells follow these steps:

1. Let's say that you have entered a formula in cell C1 that adds the contents of cells A1 and B1 and you would like to copy this formula to cells C2 and C3 so that C2 will contain the sum of A2 and B2 and cell C3 will contain the sum of A3 and B3.

2. First click in cell **C1** to make it the active cell. You will see =A1+B1 in the Formula Bar.

C1 = =A1+B1

	A	B	C	D	E	F	G
1	1	1	2				
2	2	2					
3	3	3					

3. At the top of the screen, click **Edit** and select **Copy**.

4. Highlight cells C2 and C3 by clicking and dragging over them.

	A	B	C	D	E	F	G
1	1	1	2				
2	2	2					
3	3	3					
4							

5. At the top of the screen, click **Edit** and select **Paste**. The sums should now be displayed in cells C2 and C3.

C2 = =A2+B2

	A	B	C	D	E	F	G
1	1	1	2				
2	2	2	4				
3	3	3	6				
4							

GS 4.3 Moving Information

If you would like to move the contents of one cell from one location to another in the worksheet, follow these steps:

1. Click on the cell containing the information that you would like to move.

2. At the top of the screen, click **Edit** and select **Cut**.

3. Click on the target cell where you want the information to be placed.

4. At the top of the screen, click **Edit** and select **Paste**.

If you would like to move the contents of a range of cells to a different location in the worksheet, follow these steps:

1. Click and drag over the range of cells that you would like to move so that it is highlighted.

2. At the top of the screen, click **Edit** and select **Cut**.

3. Click the topmost cell of the new location. (It is not necessary to click and drag over the entire range of the new location.)

4. At the top of the screen, click **Edit** and select **Paste**.

If you make a mistake, just click ***Edit*** *and select* ***Undo****.*

GS 4.4 Changing the Column Width

There are a couple of different ways that you can use to change the column width. Only one way will be described here. Output from the Descriptive Statistics data analysis tool will be used as an example. As you can see in the output displayed below, many of the labels in column A can only be partially viewed because the column width is too narrow.

	A	B
1	*Test Score*	
2		
3	Mean	23.89474
4	Standard E	0.752752
5	Median	24
6	Mode	21
7	Standard D	3.281171
8	Sample Va	10.76608

Position the mouse pointer directly on the vertical line between A and B in the letter row at the top of the worksheet — A | B —so that it turns into a black plus sign.

Click and drag to the right until you can read all the output labels. (You can also click and drag to the left to make columns narrower.) After adjusting the column width, your output should appear similar to the output shown below.

	A	B
1	*Test Score*	
2		
3	Mean	23.89474
4	Standard Error	0.752752
5	Median	24
6	Mode	21
7	Standard Deviation	3.281171
8	Sample Variance	10.76608

Getting Started with Sorting Information

GS 5.1 Sorting a Single Column of Information

Let's say that you have entered "Score" in cell A1 and four numbers directly below it and that you would like to sort the numbers in ascending order.

	A	B	C	D	E	F	G
1	Score						
2	15						
3	79						
4	18						
5	2						

1. Click and drag from cell A1 to cell A5 so that the range of cells is highlighted.

You could also click directly on A *in the letter row at the top of the worksheet. This will result in all cells of column A being highlighted.*

	A	B	C	D	E	F	G
1	Score						
2	15						
3	79						
4	18						
5	2						
6							

2. At the top of the screen, click **Data** and select **Sort**.

3. In the Sort dialog box that appears, you are given the choice of sorting the information in column A in either ascending or descending order. The ascending order has already been selected. Header row has also been selected. This means that the "Score" header will stay in cell A1 and will not be included in the sort. Click **OK** at the bottom of the dialog box.

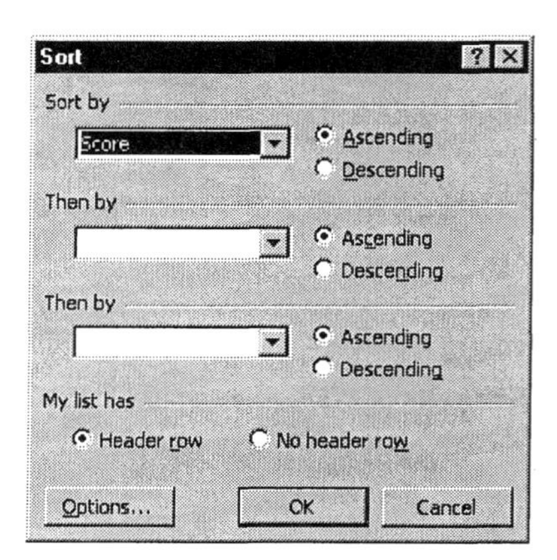

The cells in column A should now be sorted in ascending order as shown below.

	A	B	C	D	E	F	G
1	Score						
2	2						
3	15						
4	18						
5	79						
6							

GS 5.2 Sorting Multiple Columns of Information

Your Excel data files will frequently contain multiple columns of information. When you sort multiple columns at the same time, Excel provides a number of options.

Let's say that you have a data file that contains the information shown below and that you would like to sort the file by GPA in descending order.

	A	B	C	D	E	F	G
1	Score	Age	Major	GPA			
2	2	19	Music	3.1			
3	15	19	History	2.4			
4	18	22	English	2.7			
5	79	20	English	3.7			

1. Click and drag from A1 down and across to D5 so that the entire range of cells is highlighted.

	A	B	C	D	E	F	G
1	Score	Age	Major	GPA			
2	2	19	Music	3.1			
3	15	19	History	2.4			
4	18	22	English	2.7			
5	79	20	English	3.7			

2. At the top of the screen, click **Data** and select **Sort**.

3. In the Sort dialog box that appears, you are given the option of sorting the data by three different variables. You want to sort only by GPA in descending order. Click the down arrow to the right of the Sort by window until you see GPA and click on **GPA** to select it. Then click the button to the left of **Descending** so that a black dot appears there. You want the variable labels to stay in row 1, so **Header row** should be selected. Click **OK**.

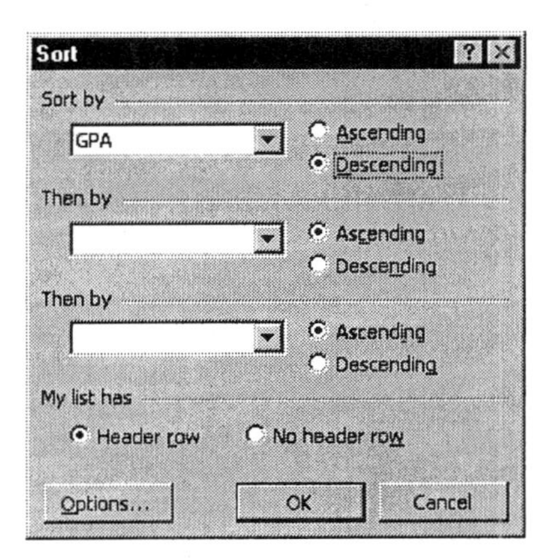

The sorted data file is shown below.

	A	B	C	D	E	F	G
1	Score	Age	Major	GPA			
2	79	20	English	3.7			
3	2	19	Music	3.1			
4	18	22	English	2.7			
5	15	19	History	2.4			

Getting Started with Saving Information

GS 6.1 Saving Files

To save a newly created file for the first time, click **File** and select **Save** at the top of the screen. A Save As dialog box will appear. You will need to select the location for saving the file by clicking on it. In the dialog box shown below, the 3½ Floppy has been selected.

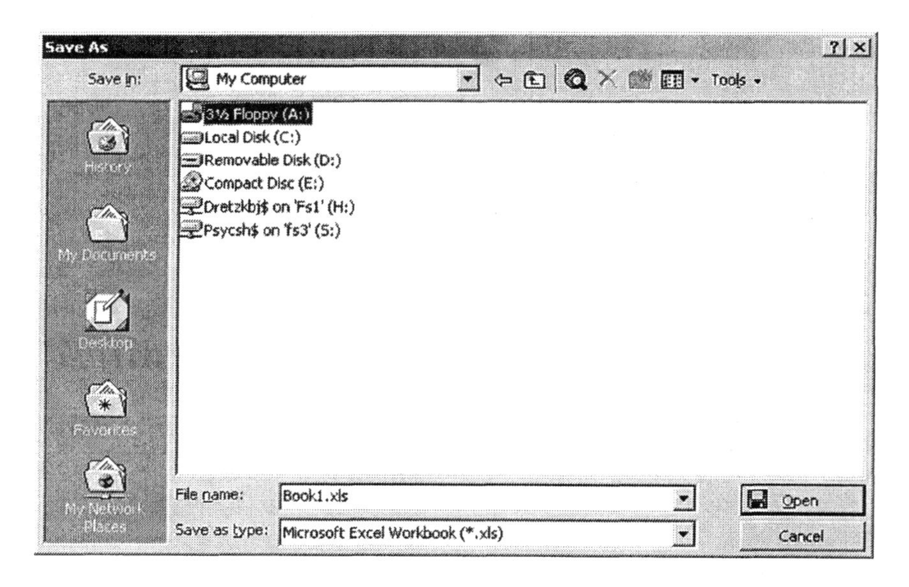

The default file name, displayed in the File name window, is **Book1.xls**. It is recommended you replace the default name with a name that is more descriptive. It is also recommended that you use the **xls** extension for all your Excel files.

Once you have saved a file, clicking **File** and selecting **Save** will result in the file being saved in the same location under the same file name. No dialog box will appear. If you would like to save the file in a different location, you will need to click **File** and select **Save As**.

GS 6.2 Naming Files

Windows 2000 and Mac versions of Excel will allow file names to have around 200 characters. The extension can have up to three characters. You will find that long, descriptive names will be easier to work with than really short names. For example, if a file contains data that was collected in a survey of Milwaukee residents, you may want to name the file **Milwaukee residents survey.xls**.

Several symbols cannot be used in file names. These include: forward slash (/), backslash (\), greater-than sign (>), less-than sign (<), asterisk (*), question mark (?), quotation mark ("), pipe symbol (|), color (:), and semicolon (;).

Getting Started with Printing Information

GS 7.1 Printing Worksheets

To print a worksheet, click **File** and select **Print**. The Print dialog box, displayed at the top of the next page, will appear.

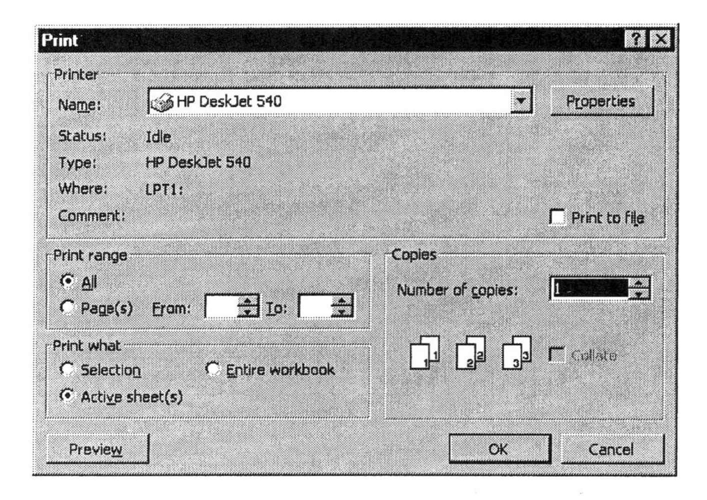

Under Print range, you will usually select **All**, and under Print what, you will usually select **Active sheet(s)**. The default number of copies is 1, but you can increase this if you need more copies. When the Print dialog box has been completed as you would like, click **OK**.

GS 7.2 Page Setup

Excel provides a number of page setup options for printing worksheets. To access these options, click **File** and select **Page Setup**. Under **Page**, you may want to select the Landscape orientation for worksheets that have several columns of data. Under **Sheet**, you may want to select Gridlines.

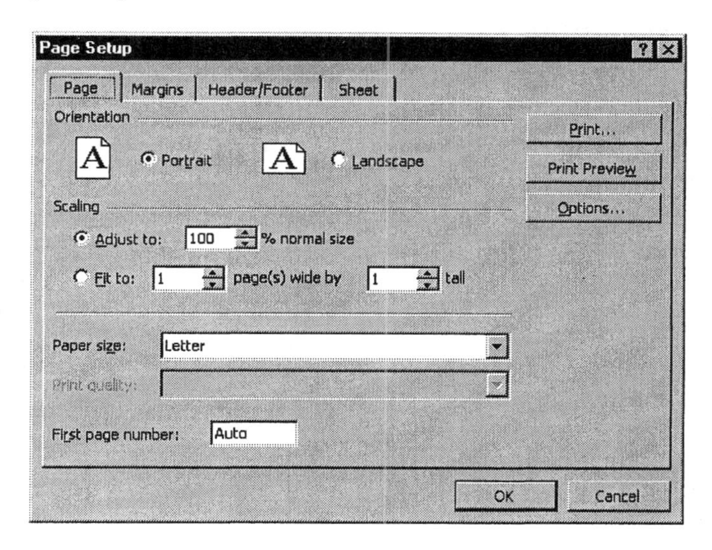

Getting Started with Add-ins

GS 8.1 Loading Excel's Analysis Toolpak

The Analysis Tookpak is an Excel Add-In that may not necessarily be loaded. If it does not appear at the bottom of the Tools menu, then click on **Add-Ins** in the Tools menu to get the dialog box shown at the top of the next page.

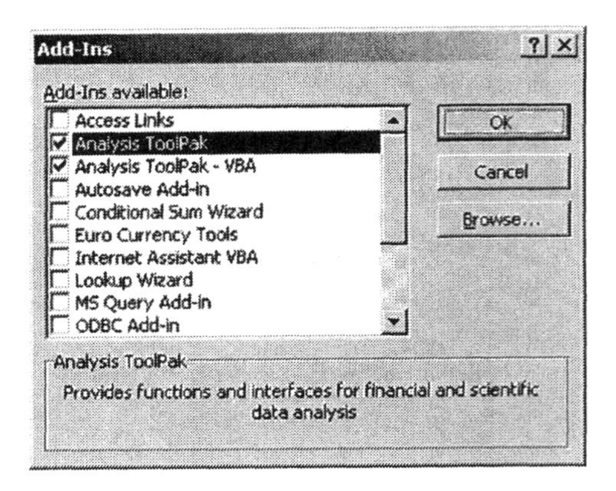

Click in the box to the left of **Analysis ToolPak** to place a checkmark there. Then click **OK**. The ToolPak will load and will be listed at the bottom of the Tools menu as shown below.

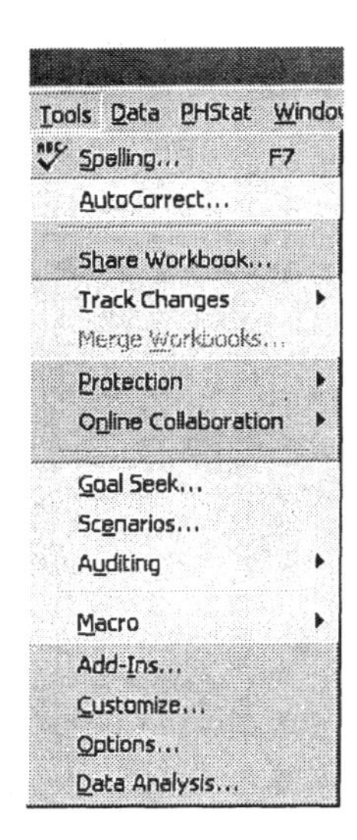

GS 8.2 Loading the PHStat2 Add-In

PHStat2 is a Prentice Hall statistical add-in that is included on the CD-ROM that accompanies your statistics textbook. PHStat2 is Windows software. The instructions that are given here also appear in the PHStat2 readme file. Read through that file completely to make sure that you are aware of all the technical requirements.

To use the Prentice Hall PHStat2 Microsoft Excel add-in, you first need to run the setup program (Setup.exe) located in the PHStat2 folder on your textbook CD-ROM. The setup program will install the PHStat2 program files to your system and add icons on your Desktop and Start Menu for PHStat2. Depending on the age of your Windows system files, some Windows system files may be updated during the setup process as well. Note that PHStat2 is compatible with Microsoft Excel 97 and Microsoft Excel 2000. PHStat2 is not compatible with Microsoft Excel 95.

During the Setup program you will have the opportunity to specify the directory into which to copy the PHStat2 files (default is \Program Files\Prentice Hall\PHStat2).

To begin using the Prentice Hall PHStat2 Microsoft Excel add-in, click the appropriate Start Menu or Desktop icon for PHStat2 that was added to your system during the setup process.

When a new, blank Excel worksheet appears, check the Tools menu to make sure that both the **Analysis ToolPak** and **Analysis ToolPak–VBA** have been checked.

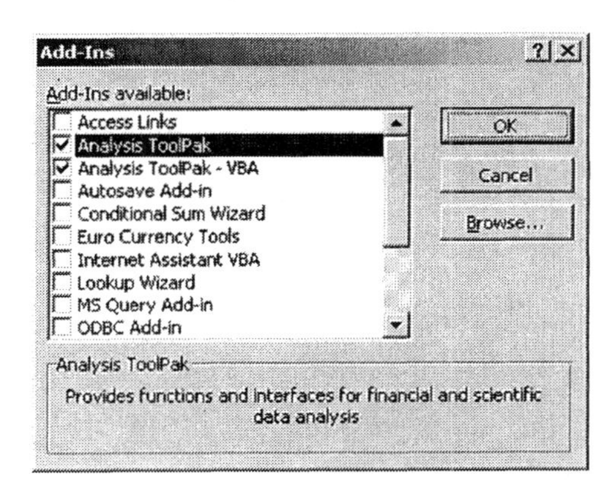

Introduction to Statistics

CHAPTER 1

1.2 Observational Studies; Simple Random Sampling

► Example 3 (pg. 16) Obtaining a Simple Random Sample

You will be generating a sample of size 10 from the 8,791 residents of Lemont. The problem is displayed in Example 3 on page 16. To obtain the random sample, you will be using the PHStat add-in.

If the PHStat add-in has not been loaded, you will need to load it before continuing. Follow the instructions in Section GS 8.2.

---

1. Open a new, blank Excel worksheet.

2. At the top of the screen, click **PHStat**. Select **Sampling → Random Sample Generation**.

3. Complete the Random Sample Generator dialog box as shown below. A sample of 10 residents will be randomly selected from a population of 8,791. The topmost cell of the output will contain the label "Resident #." Click **OK**.

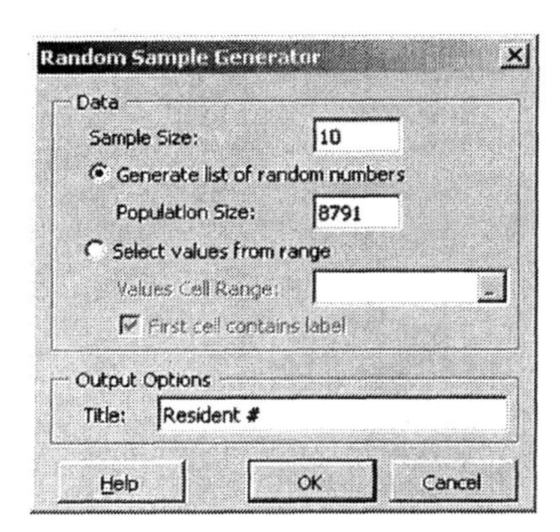

The output is displayed in a new worksheet named "RandomNumbers." Because the numbers were generated randomly, it is not likely that your output will be exactly the same.

	A	B	C	D	E	F	G
1	Resident #						
2	5725						
3	2415						
4	5463						
5	3224						
6	4049						
7	1112						
8	4166						
9	7321						
10	2084						
11	4559						

◀

► Exercise 15 (pg. 19) Obtaining Two Simple Random Samples of States

You will be generating two lists of 10 random numbers between 1 and 50 to use in selecting random samples of states.

If the PHStat add-in has not been loaded, you will need to load it before continuing. Follow the instructions in Section GS 8.2.

1. Open a new, blank Excel worksheet.

2. At the top of the screen, click **PHStat**. Select **Sampling → Random Sample Generation.**

3. Complete the Random Sample Generator dialog box as shown below. A sample of 10 states will be randomly selected from a population of 50. "State #" will appear in the top cell of the output. Click **OK.**

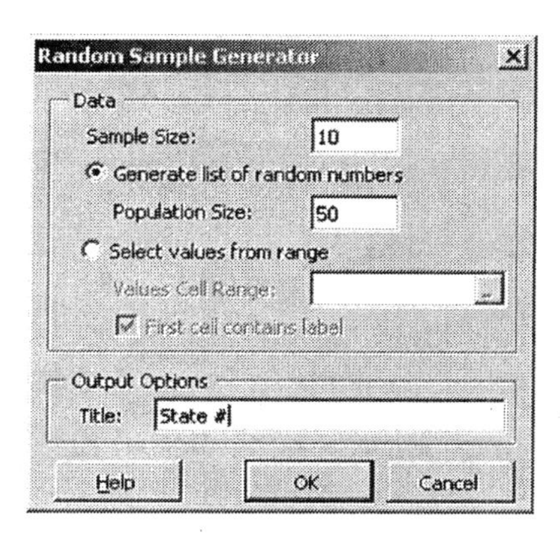

The ten randomly selected states are displayed below. Because the numbers were generated randomly, it is not likely that your output will be exactly the same.

	A	B	C	D	E	F	G
1	State #						
2	13						
3	36						
4	22						
5	50						
6	27						
7	47						
8	21						
9	23						
10	10						
11	14						

4. Repeat the procedure to obtain a second sample of 10 states. At the top of the screen, click **PHStat**. Select **Sampling → Random Sample Generation.**

5. Complete the Random Sample Generator dialog box as shown below. A second sample of 10 states will be randomly selected. Click **OK**.

The output is placed in a worksheet named "RandomNumbers2." The ten randomly selected states are displayed below. Again, because the numbers were generated randomly, it is not likely that your output will be exactly the same.

	A	B	C	D	E	F	G
1	Randomly Selected Values						
2	14						
3	41						
4	37						
5	30						
6	34						
7	11						
8	20						
9	3						
10	43						
11	18						

◀

► Exercise 16 (pg. 20) Obtaining a Simple Random Sample of Presidents

If the PHStat add-in has not been loaded, you will need to load it before continuing. Follow the instructions in Section GS 8.2.

1. Open a new Excel worksheet.

2. At the top of the screen, click **PHStat**. Select **Sampling → Random Sample Generation**.

3. Complete the Random Sample Generator dialog box as shown below. A sample of eight presidents will be randomly selected from a population of 43. "President #" will appear in the top cell of the generated output. Click **OK**.

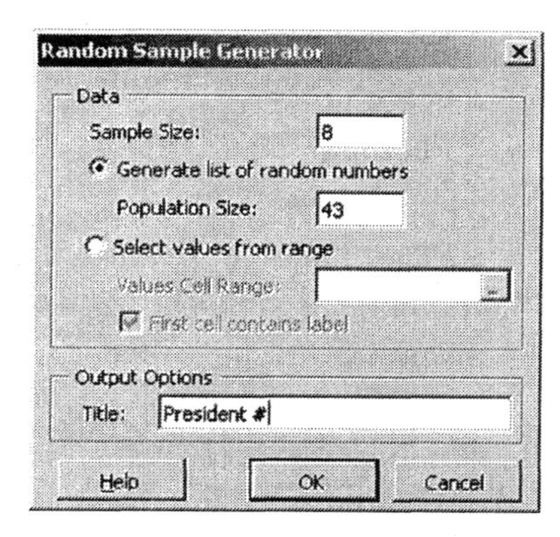

The set of eight random numbers is displayed below. Because the numbers were generated randomly, it is unlikely that your output will be exactly the same.

	A	B	C	D	E	F	G
1	President #						
2	6						
3	20						
4	13						
5	27						
6	22						
7	37						
8	40						
9	36						

4. Repeat the procedure to obtain a second sample of eight presidents. At the top of the screen, click **PHStat**. Select **Sampling → Random Sample Generation**.

5. Complete the Random Sample Generator dialog box as shown below. A second random sample of eight presidents will be selected. Click **OK**.

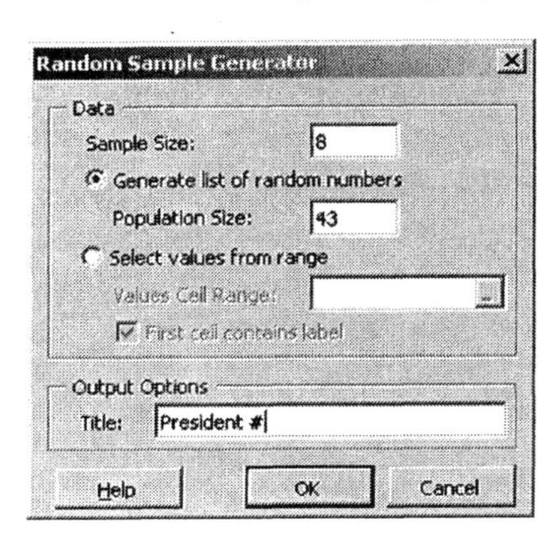

The second set of eight randomly selected presidents is displayed below. Again, because the numbers were generated randomly, it is unlikely that your output will be exactly the same.

	A	B	C	D	E	F	G
1	President #						
2	4						
3	24						
4	7						
5	27						
6	31						
7	17						
8	25						
9	34						

◀

Section 1.3 Other Types of Sampling

Obtaining a Stratified Sample

You will be generating a sample of size 100 comprised of 28 resident students, 61 non-resident students, and 11 staff.

If the PHStat add-in has not been loaded, you will need to load it before continuing. Follow the instructions in Section GS 8.2.

---

1. Open a new, blank Excel worksheet.

2. You will first select the resident students. At the top of the screen, click **PHStat**. Select **Sampling → Random Sample Generation**.

3. Complete the Random Sample Generator dialog box as shown below. A sample of 28 resident students will be randomly selected from a population of 6,204. "Resident Student #" will appear in the top cell of the generated output. Click **OK**.

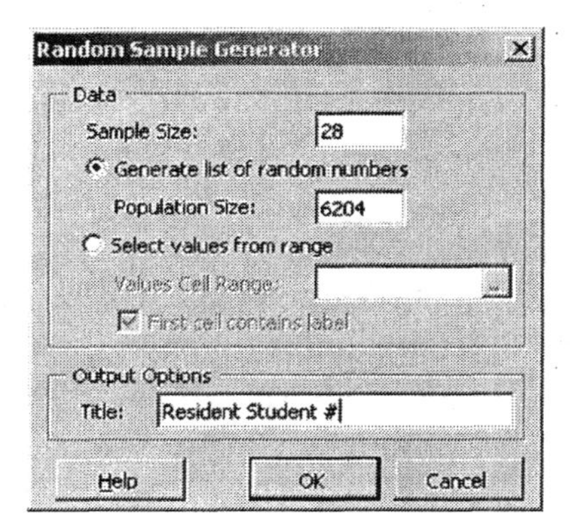

A partial listing of the 28 randomly selected resident students is displayed below. The output appears in a sheet labeled "RandomNumbers." Because the numbers were generated randomly, it is unlikely that your output will be exactly the same.

	A	B	C	D	E	F	G
1	Resident Student #						
2	3623						
3	5379						
4	4171						

4. Repeat the procedure to randomly select 61 non-resident students. At the top of the screen, click **PHStat**. Select **Sampling → Random Sample Generation**.

5. Complete the Random Sample Generator dialog box as shown below. A sample of 61 nonresident students will be selected from a population of 13,304. "Nonresident Student #" will appear in the top cell of the generated output. Click **OK**.

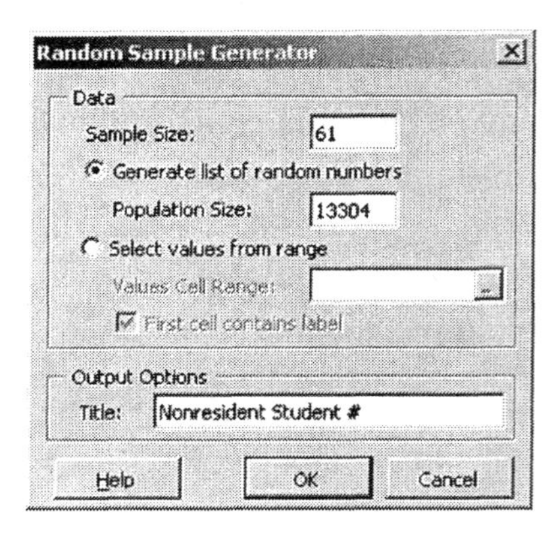

A partial listing of the 61 randomly selected nonresident students is displayed below. The output appears in a sheet labeled "RandomNumbers2." Again, because the numbers were generated randomly, it is unlikely that your output will be exactly the same.

	A	B	C	D	E	F	G
1	Nonresident Student #						
2	5125						
3	12814						
4	11759						

6. Repeat the procedure to randomly select the 11 staff members. At the top of the screen, click **PHStat**. Select **Sampling → Random Sample Generation**.

7. Complete the Random Sample Generator dialog box as shown below. A sample of 11 staff will be randomly selected from a population of 2,401. "Staff #" will appear in the top cell of the generated output. Click **OK**.

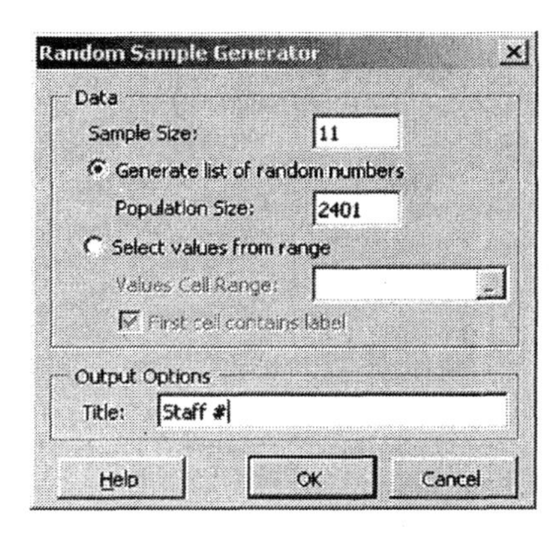

The 11 randomly selected numbers are displayed below. The output appears in a sheet labeled "RandomNumbers3." Again, because the numbers were generated randomly, it is unlikely that your output will be exactly the same.

	A	B	C	D	E	F	G
1	Staff #						
2	202						
3	393						
4	940						
5	1784						
6	642						
7	2142						
8	1463						
9	268						
10	814						
11	954						
12	431						

◀

Organizing and Summarizing Data

CHAPTER 2

Section 2.1 Organizing Qualitative Data

► Exercise 15 (pg. 64)

Constructing a Bar Graph

You will use the males' Educational Attainment data in Exercise 15 to learn how to construct a bar graph.

1. Open worksheet "2_1_15" in the Chapter 2 folder. The first few rows are shown below.

	A	B	C	D	E	F	G
1	Education:	Males (in r	Females (in millions)				
2	Not a high	16.6	16.6				
3	High schoo	31.8	34.8				
4	Some colle	17.1	17.5				

2. Click in any cell of the table. At the top of the screen, click **Insert** and select **Chart**.

If you activate a cell containing data prior to selecting Chart, the data range will be automatically entered in the chart dialog box.

3. In the Chart Type dialog box, click on the **Column** Chart type to select it. Under Chart sub-type, select the leftmost diagram in the top row. Click **Next>**.

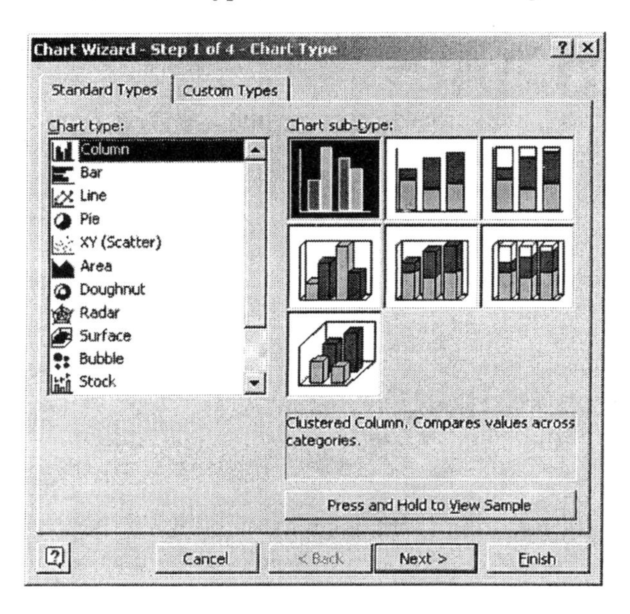

4. For this example, you will be graphing only the males' data. So, in the Data range window, change C7 to **B7** so that the females' data are not included. Click **Next>**.

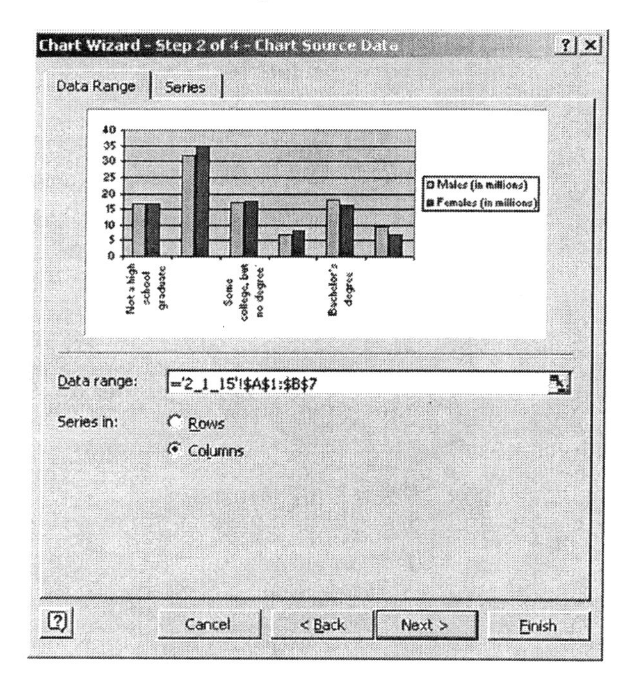

5. Click on the **Titles** tab at the top of the Chart Options dialog box. For the Chart title, enter **Educational Attainment of Males in 1999**. For the Category (X) axis, enter **Educational Attainment**. For the Value (Y) axis, enter **Frequency (in millions)**. Click **Next>**.

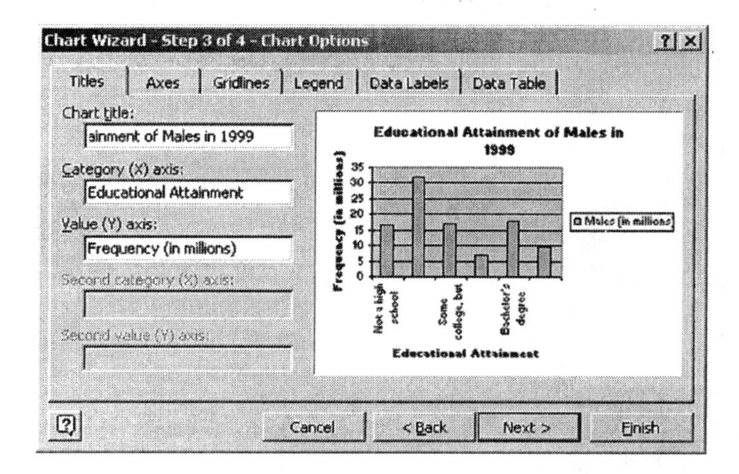

6. The Chart Location dialog box presents two options for placement of the chart. For this example, select **As new sheet**. Click **Finish**.

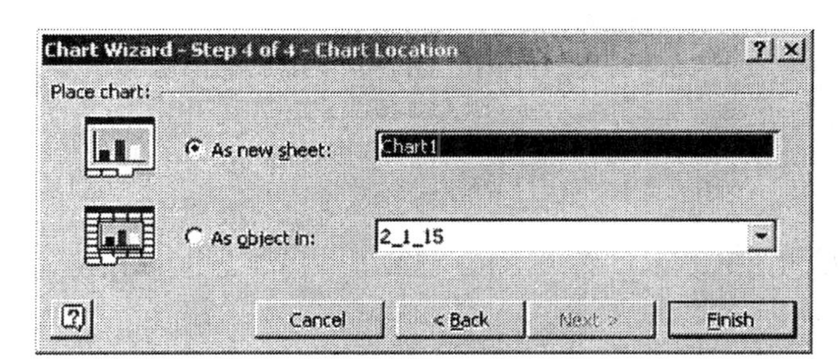

7. Your completed graph should look similar to the one shown below. If you would like to remove the legend shown to the right of the graph, click on it and then press **[Delete]**.

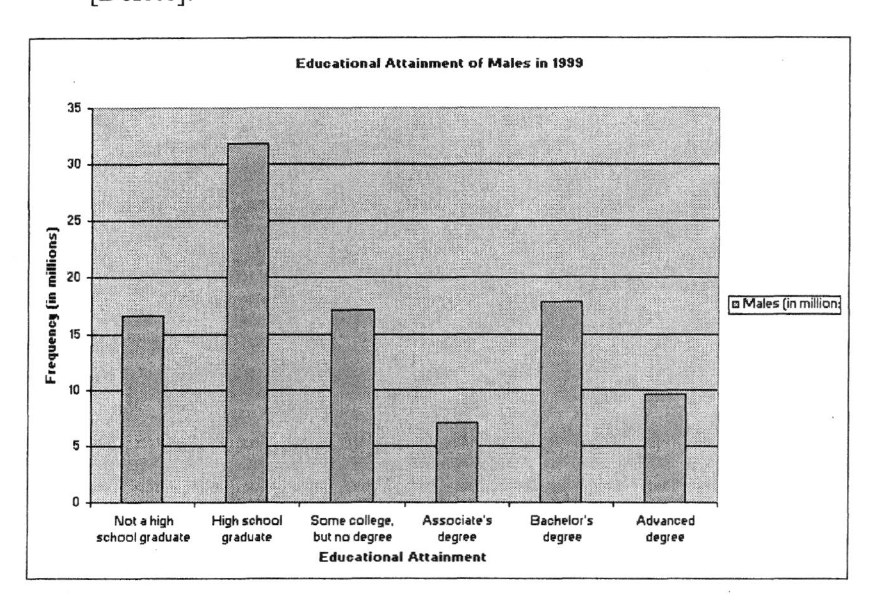

► Exercise 15 (pg. 64) Constructing a Pareto Chart

You will use the males' Educational Attainment data in Exercise15 to learn how to construct a pareto chart.

1. Open worksheet "2_1_15" in the Chapter 2 folder. The first few rows are shown below.

	A	B	C	D	E	F	G
1	Educationa	Males (in r	Females (in millions)				
2	Not a high	16.6	16.6				
3	High schoo	31.8	34.8				
4	Some colle	17.1	17.5				

2. In a pareto chart, the bars are displayed in decreasing order of frequency. So, the first thing you need to do is sort the data by Males (in millions) in descending order. The sorted data are shown below.

For instructions on how to sort, refer to Sections GS 5.1 and GS 5.2.

	A	B	C	D	E	F	G
1	Education	Males (in r	Females (in millions)				
2	High schoo	31.8	34.8				
3	Bachelor's	17.9	16.2				
4	Some colle	17.1	17.5				
5	Not a high	16.6	16.6				
6	Advanced	9.6	7				
7	Associate'	7	8				

3. Click in any cell of the sorted data table. At the top of the screen, click **Insert** and select **Chart**.

4. In the Chart Type dialog box, click on the **Column** Chart type to select it. Under Chart sub-type, select the leftmost diagram in the top row. Click **Next>**.

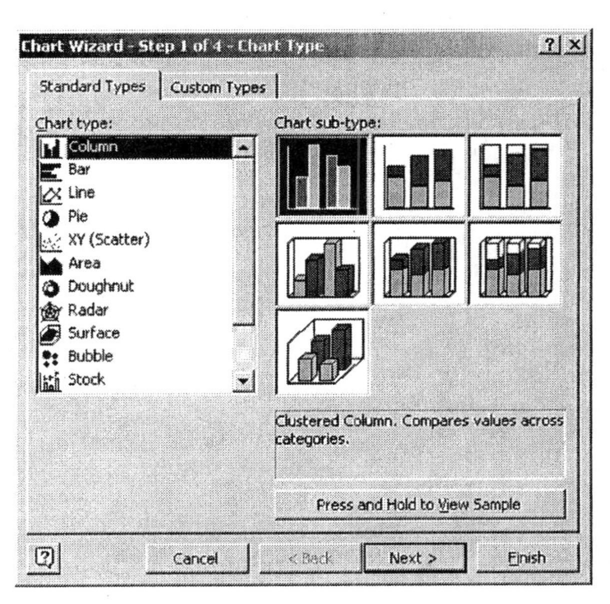

5. You will be graphing only the males' data. So, in the Data range window, change C7 to **B7** so that the females data are not included. Click **Next>**.

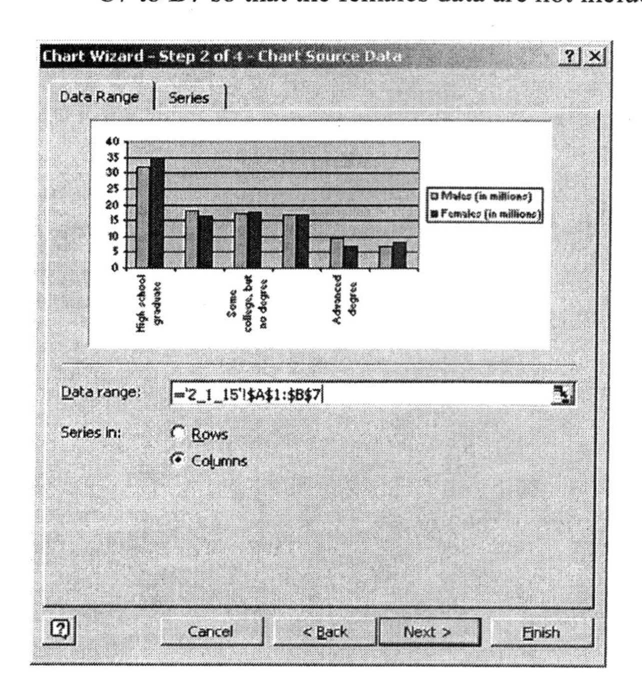

6. Click the **Titles** tab at the top of the Chart Options dialog box. For the Chart title, enter **Educational Attainment of Males in 1999**. For the Category (X) axis, enter **Educational Attainment**. For the Value (Y) axis, enter **Frequency (in millions)**.

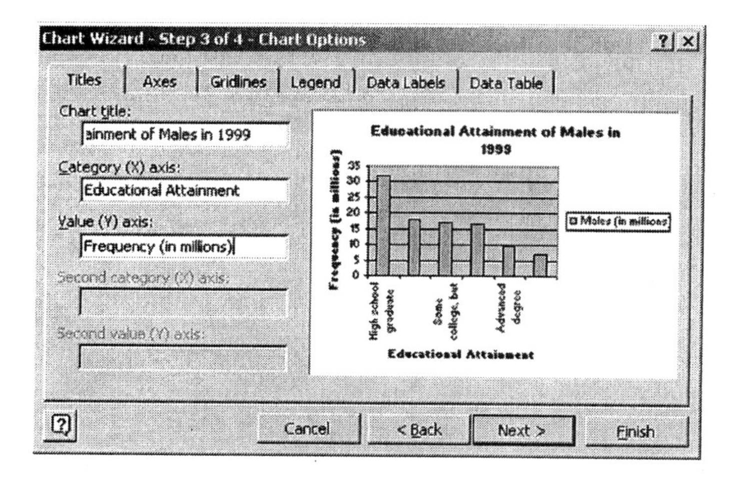

7. To remove the legend to the right of the graph, first click the **Legend** tab at the top of the dialog box. Then click in the box to the left of **Show Legend** to remove the checkmark that appears there. Click **Next>**.

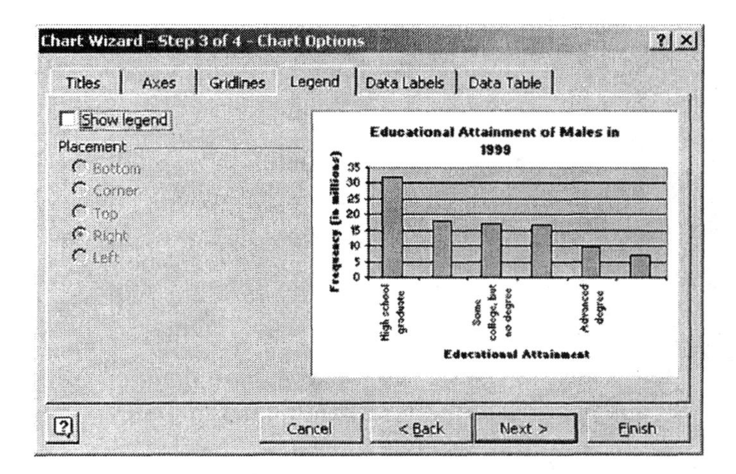

8. The Chart Location dialog box presents two options for placement of the chart. For this example, select **As new sheet**. Click **Finish**.

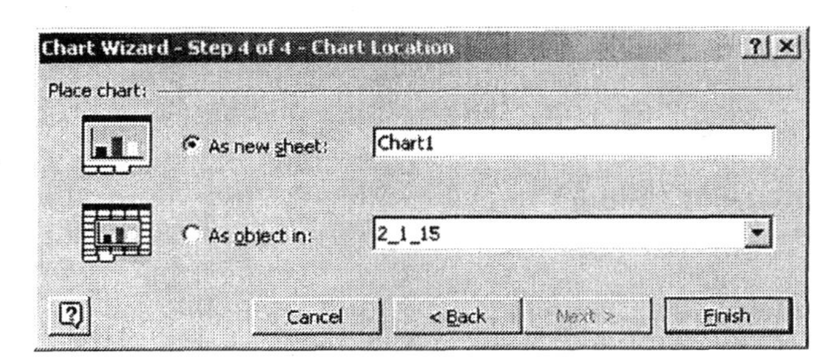

Your completed pareto chart should look similar to the one shown below.

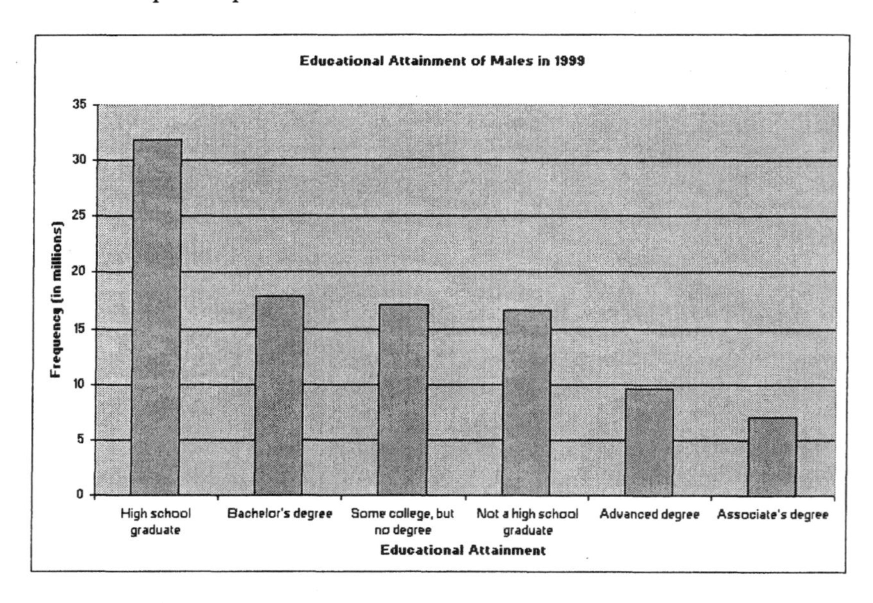

▶ Exercise 15 (pg. 64) Constructing a Pie Chart

You will use the males' Educational Attainment data in Exercise 15 to learn how to construct a pie chart.

1. Open worksheet "2_1_15" in the Chapter 2 folder. The first few rows are shown below.

	A	B	C	D	E	F	G
1	Education	Males (in r	Females (in millions)				
2	Not a high	16.6	16.6				
3	High schoo	31.8	34.8				
4	Some colle	17.1	17.5				

2. Click in any cell of the table. At the top of the screen, click **Insert** and select **Chart**.

3. In the Chart Type dialog box, click on the **Pie** Chart type to select it. Under Chart sub-type, select the leftmost diagram in the top row. Click **Next>**.

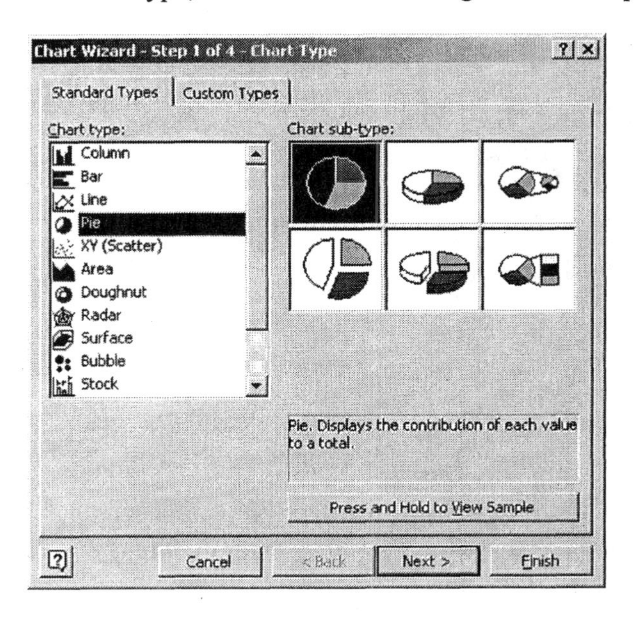

4. The Data range in the Chart Source Data dialog box should be the same as the range shown below. Make any necessary corrections. Click **Next>**.

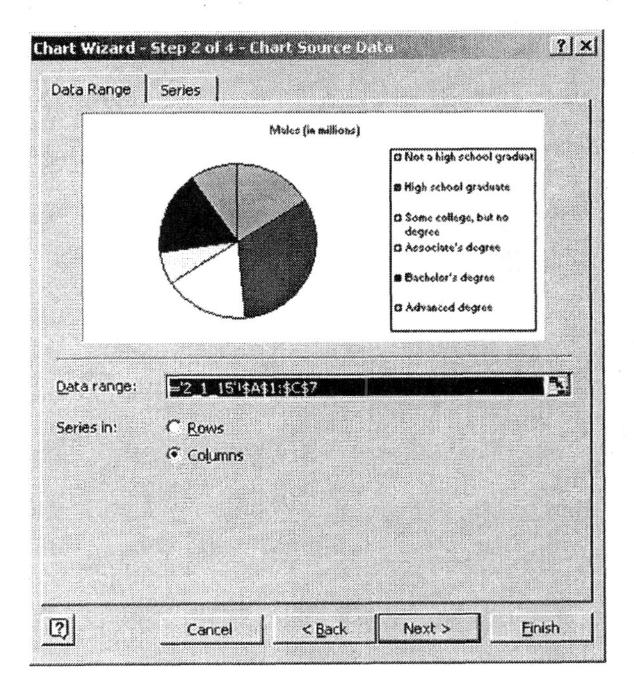

5. Click the **Titles** tab at the top of the Chart Options dialog box. For the Chart title, enter **Educational Attainment of Males in 1999**.

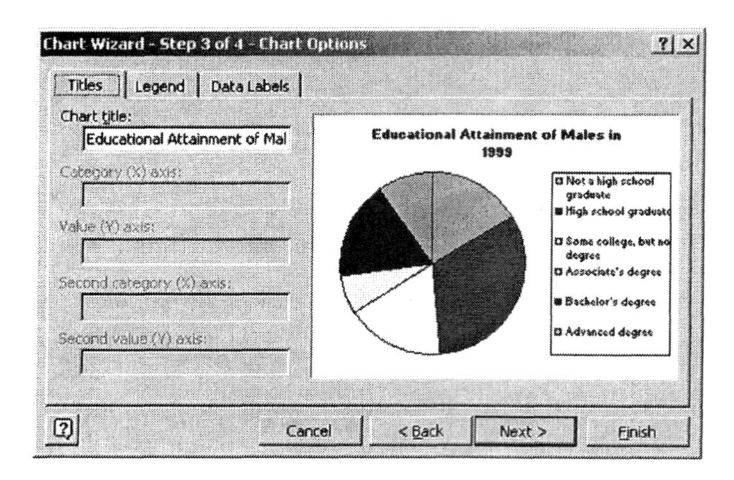

6. Click the **Legend** tab at the top of the Chart Options dialog box. You want a legend to show on your completed pie chart, so there should be a checkmark in the box to the left of **Show Legend**. If a checkmark does not appear, you can place one there by clicking in the box. Several placement options are provided. For this example, click in the button next to **Left** to select it.

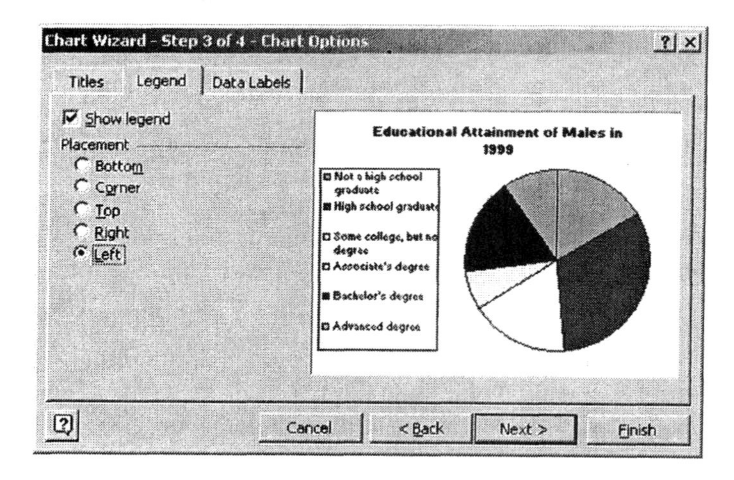

7. Click on the **Data Labels** tab at the top of the Chart Options dialog box. Under Data labels, select **Show percent** by clicking in the button. Click **Next>**.

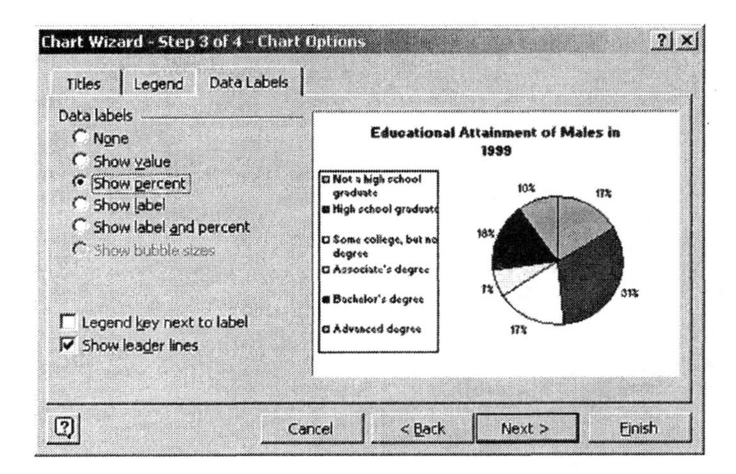

8. The Chart Location dialog box presents two options for placement of the chart. For this example, select **As new sheet**. Click **Finish**.

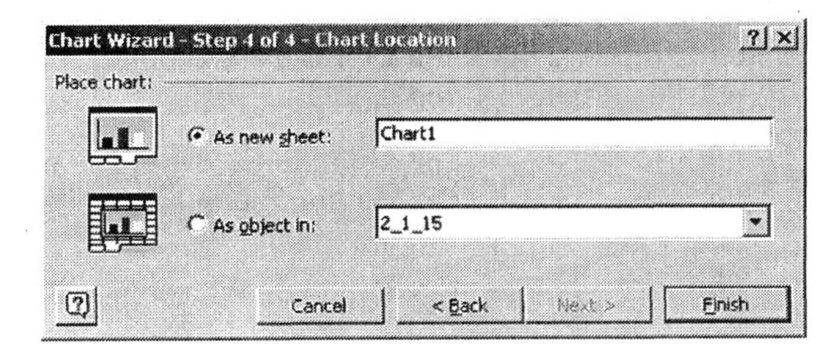

9. Some of the labels in the legend do not appear in their entirety because the box around the legend is too small. To adjust the box size, click anywhere in the legend box. Black handles will appear. Move the cursor so that it is directly on top of a handle. Then click and drag the border to make the box larger.

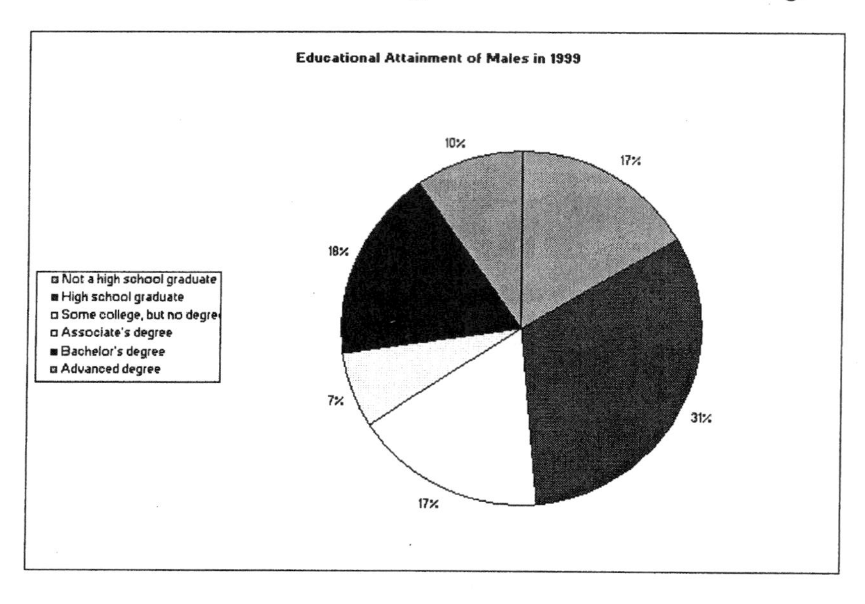

The pie chart with complete labels in the legend is shown below.

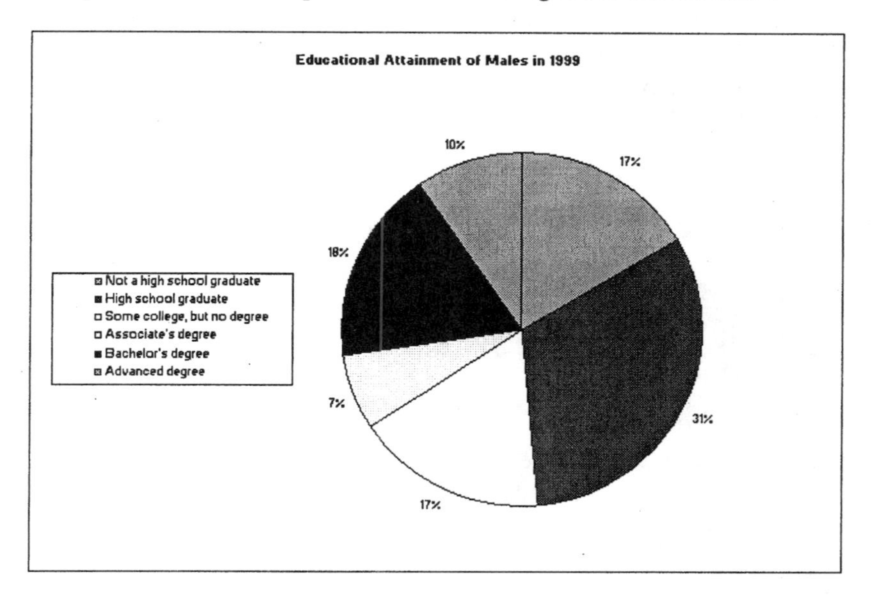

◄

► Exercise 15 (pg. 64) Constructing a Side-by-Side Relative Frequency Bar Graph

You will first construct a relative frequency distribution of educational attainment of males and females in the United States. You will then construct a side-by-side relative frequency bar graph of the same data.

1. Open worksheet "2_1_15" in the Chapter 2 folder. The first few rows are shown below.

	A	B	C	D	E	F	G
1	Educationa	Males (in r	Females (in millions)				
2	Not a high	16.6	16.6				
3	High schoo	31.8	34.8				
4	Some colle	17.1	17.5				

2. To calculate the sum of the male frequencies, click in cell **B8** and then click the **AutoSum** button at the top of the screen. The AutoSum button is a capital Greek sigma (Σ). The SUM function will sum the numbers in the range shown in parentheses. That range should be B2:B7. If your range is different, enter the correct range and then press [**Enter**].

	A	B	C	D	E	F	G
1	Educationa	Males (in r	Females (in millions)				
2	Not a high	16.6	16.6				
3	High scho	31.8	34.8				
4	Some coll	17.1	17.5				
5	Associate'	7	8				
6	Bachelor's	17.9	16.2				
7	Advanced	9.6	7				
8		=SUM(B2:B7)					

3. To sum the females' frequencies, click in cell **C8,** click the **AutoSum** button Σ at the top of the screen, and check to be sure that the range is C2:C7.

	A	B	C	D	E	F	G
1	Educationa	Males (in r	Females (in millions)				
2	Not a high	16.6	16.6				
3	High schoo	31.8	34.8				
4	Some colle	17.1	17.5				
5	Associate'	7	8				
6	Bachelor's	17.9	16.2				
7	Advanced	9.6	7				
8		100	=SUM(C2:C7)				

4. Make any necessary corrections to the range and then press [**Enter**].

	A	B	C	D	E	F	G
1	Educationa	Males (in r	Females (in millions)				
2	Not a high	16.6	16.6				
3	High schoo	31.8	34.8				
4	Some colle	17.1	17.5				
5	Associate'	7	8				
6	Bachelor's	17.9	16.2				
7	Advanced	9.6	7				
8		100	100.1				
9							

5. Copy the data to an area below the table so that you can calculate relative frequencies. To do this, click and drag over the entire table so that it is highlighted.

	A	B	C	D	E	F	G
1	Education	Males (in r	Females (in millions)				
2	Not a high	16.6	16.6				
3	High schoo	31.8	34.8				
4	Some colle	17.1	17.5				
5	Associate'	7	8				
6	Bachelor's	17.9	16.2				
7	Advanced	9.6	7				
8		100	100.1				

6. Click the **Copy** button. Then, to place the copy below the original table, click in cell **A10** and then click the **Paste** button.

	A	B	C	D	E	F	G
1	Education	Males (in r	Females (in millions)				
2	Not a high	16.6	16.6				
3	High schoo	31.8	34.8				
4	Some colle	17.1	17.5				
5	Associate'	7	8				
6	Bachelor's	17.9	16.2				
7	Advanced	9.6	7				
8		100	100.1				
9							
10	Education	Males (in r	Females (in millions)				
11	Not a high	16.6	16.6				
12	High schoo	31.8	34.8				
13	Some colle	17.1	17.5				
14	Associate'	7	8				
15	Bachelor's	17.9	16.2				
16	Advanced	9.6	7				
17		100	100.1				

7. You will now compute percentages beginning with males who are in the "Not a high school graduate" category. Click in cell **B11**.

8. To compute the percentage, you will divide the frequency by the males' total and then multiply by 100. Enter the formula **=B2/$B$8*100** as shown below. Press **[Enter]**.

The dollar signs must appear in the cell B8 reference but not in the cell B2 reference. The dollar signs make the cell address absolute so that it will not change when it is copied.

	A	B	C	D	E	F	G
1	Education	Males (in r	Females (in millions)				
2	Not a high	16.6	16.6				
3	High schoo	31.8	34.8				
4	Some colle	17.1	17.5				
5	Associate'	7	8				
6	Bachelor's	17.9	16.2				
7	Advanced	9.6	7				
8		100	100.1				
9							
10	Education	Males (in r	Females (in millions)				
11		=B2/B8*100	16.6				
12	High schoo	31.8	34.8				
13	Some colle	17.1	17.5				
14	Associate'	7	8				
15	Bachelor's	17.9	16.2				
16	Advanced	9.6	7				
17		100	100.1				

9. Copy the contents of cell B11 to cells B12 through B17.

For this problem, the percentage values in B11:B17 are exactly the same as the frequency values in B2:B8, because the sum of the frequencies is 100.

	A	B	C	D	E	F	G
1	Education	Males (in r	Females (in millions)				
2	Not a high	16.6	16.6				
3	High schoo	31.8	34.8				
4	Some colle	17.1	17.5				
5	Associate'	7	8				
6	Bachelor's	17.9	16.2				
7	Advanced	9.6	7				
8		100	100.1				
9							
10	Education	Males (in r	Females (in millions)				
11	Not a high	16.6	16.6				
12	High scho	31.8	34.8				
13	Some coll	17.1	17.5				
14	Associate	7	8				
15	Bachelor's	17.9	16.2				
16	Advanced	9.6	7				
17		100	100.1				

10. You will now compute percentages for females beginning with females classified as "Not a high school graduate." Click in cell **C11**.

	A	B	C	D	E	F	G
1	Educationa	Males (in r	Females (in millions)				
2	Not a high	16.6	16.6				
3	High schoo	31.8	34.8				
4	Some colle	17.1	17.5				
5	Associate'	7	8				
6	Bachelor's	17.9	16.2				
7	Advanced	9.6	7				
8		100	100.1				
9							
10	Educationa	Males (in r	Females (in millions)				
11	Not a high	16.6	16.6				
12	High schoo	31.8	34.8				
13	Some colle	17.1	17.5				
14	Associate'	7	8				
15	Bachelor's	17.9	16.2				
16	Advanced	9.6	7				
17		100	100.1				

11. To compute the percentage, you will divide the frequency by the females' total and then multiply by 100. Enter the formula **=C2/$C$8*100** as shown below. Press **[Enter]**.

To enter a cell address into a formula, you can key it in. A faster way is to use the mouse to click in the cell.

	A	B	C	D	E	F	G
1	Educationa	Males (in r	Females (in millions)				
2	Not a high	16.6	16.6				
3	High schoo	31.8	34.8				
4	Some colle	17.1	17.5				
5	Associate'	7	8				
6	Bachelor's	17.9	16.2				
7	Advanced	9.6	7				
8		100	100.1				
9							
10	Educationa	Males (in r	Females (in millions)				
11	Not a high		=C2/$C$8*100				
12	High schoo	31.8	34.8				
13	Some colle	17.1	17.5				
14	Associate'	7	8				
15	Bachelor's	17.9	16.2				
16	Advanced	9.6	7				
17		100	100.1				

12. Copy the contents of cell C11 to cells C12 through C17.

	A	B	C	D	E	F	G
1	Education	Males (in r	Females (in millions)				
2	Not a high	16.6	16.6				
3	High schoo	31.8	34.8				
4	Some colle	17.1	17.5				
5	Associate'	7	8				
6	Bachelor's	17.9	16.2				
7	Advanced	9.6	7				
8		100	100.1				
9							
10	Education	Males (in r	Females (in millions)				
11	Not a high	16.6	16.58342				
12	High schoo	31.8	34.76523				
13	Some colle	17.1	17.48252				
14	Associate'	7	7.992008				
15	Bachelor's	17.9	16.18382				
16	Advanced	9.6	6.993007				
17		100	100				

13. Because these are percentages rather than frequencies, you want to remove "(in millions)" from the column labels. First click in cell **B10** and revise the label so that it reads **Males**. Then click in cell **Cl0** and revise the label so that it reads **Females**.

	A	B	C	D	E	F	G
1	Education	Males (in r	Females (in millions)				
2	Not a high	16.6	16.6				
3	High schoo	31.8	34.8				
4	Some colle	17.1	17.5				
5	Associate'	7	8				
6	Bachelor's	17.9	16.2				
7	Advanced	9.6	7				
8		100	100.1				
9							
10	Education	Males	Females				
11	Not a high	16.6	16.58342				
12	High schoo	31.8	34.76523				
13	Some colle	17.1	17.48252				
14	Associate'	7	7.992008				
15	Bachelor's	17.9	16.18382				
16	Advanced	9.6	6.993007				
17		100	100				

14. The relative frequency distribution is now complete, and you are ready to start the side-by-side relative frequency bar graph. Click in any cell in the percentages table. Then, at the top of the screen, click **Insert** and select **Chart**.

If you activate a cell containing data prior to selecting Chart, the data range will be automatically entered in the chart dialog box.

---

15. In Chart Type dialog box, click on the **Column** Chart type to select it. Under Chart sub-type, select the leftmost diagram in the top row. Click **Next>**.

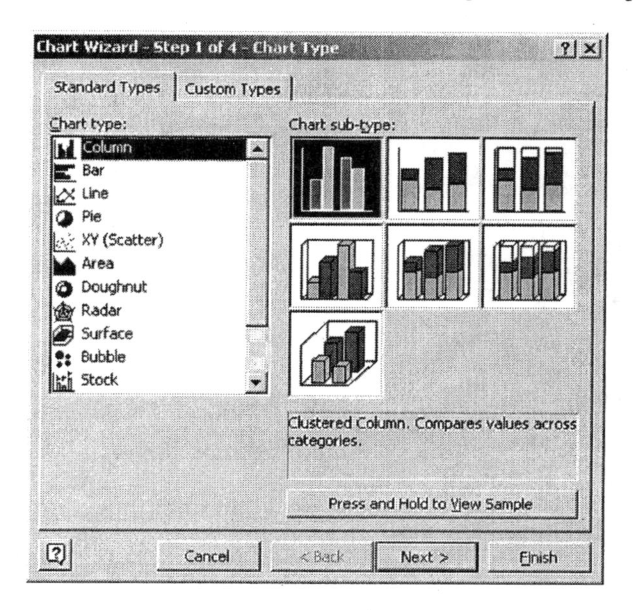

16. In the Data range window, change the last cell address from C17 to **C16** so that the totals are not included in the graph. Click **Next>**.

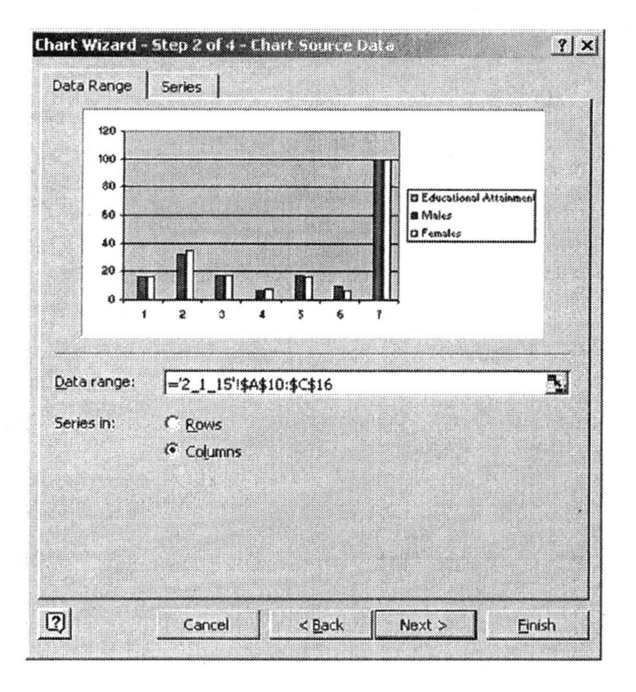

17. Click the **Titles** tab at the top of the Chart Options dialog box. For the Chart title, enter **Educational Attainment of Males versus Females in 1999**. For the Category (X) axis, enter **Educational Attainment**. For the Value (Y) axis, enter **Percentage**. Click **Next>**.

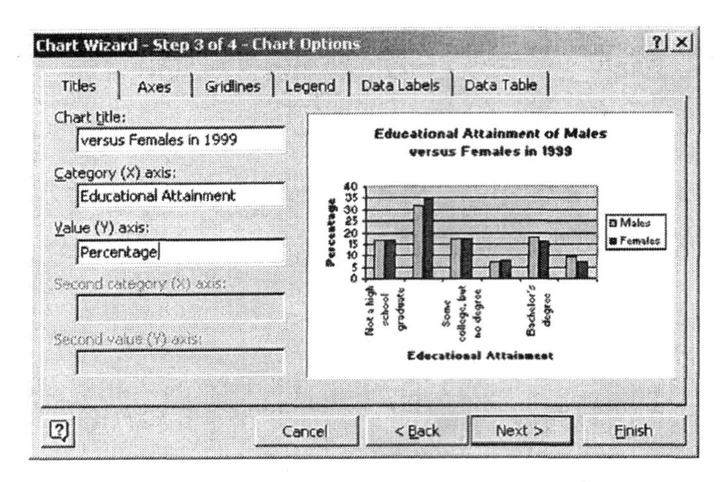

18. The Chart Location dialog box presents two options for placement of the chart. For this example, select **As new sheet**. Click **Finish**.

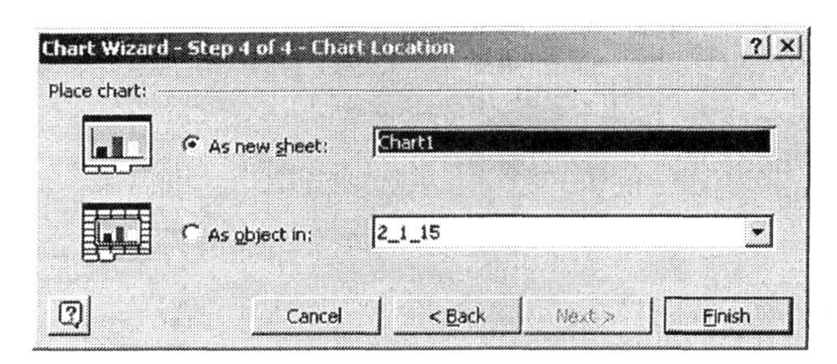

Your graph should look similar to the one shown below.

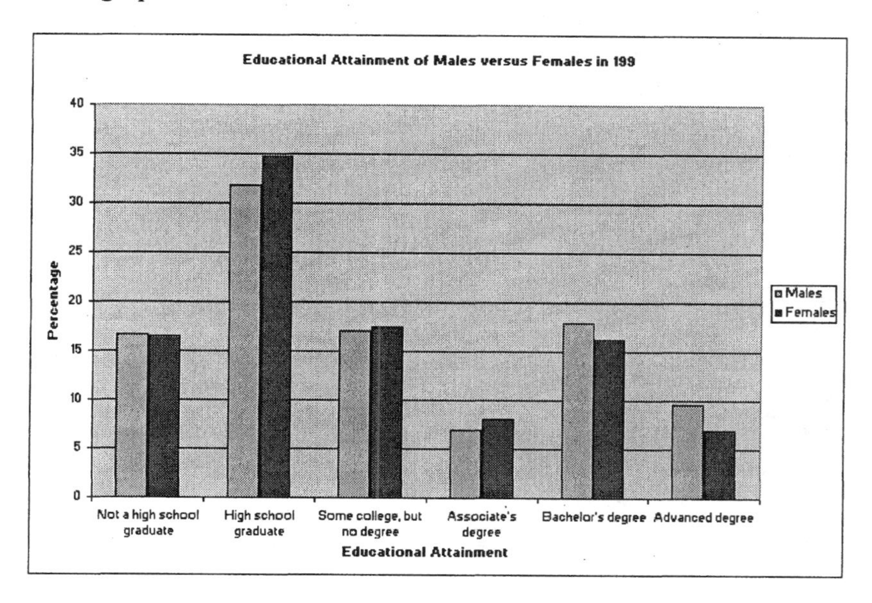

► Exercise 19 (pg. 65) Constructing a Frequency Distribution, Bar Graph, and Pie Chart

You will use the 2000 Presidential Election data to construct a frequency distribution, relative frequency distribution, frequency bar graph, and pie chart.

1. Open worksheet "2_1_19" in the Chapter 2 folder. The first few rows are shown below.

	A	B	C	D	E	F	G
1	C1-T						
2	Bush						
3	Gore						
4	Gore						

2. Click in any cell in the column of data. Then, at the top of the screen, click **Data** and select **Pivot Table and Pivot Chart Report**.

3. At the top of the dialog box, select **Microsoft Excel list or database**. At the bottom of the dialog box, select **Pivot Chart (with Pivot Table)**. Excel will prepare both a frequency distribution and a frequency bar graph of the presidential election data. Click **Next>**.

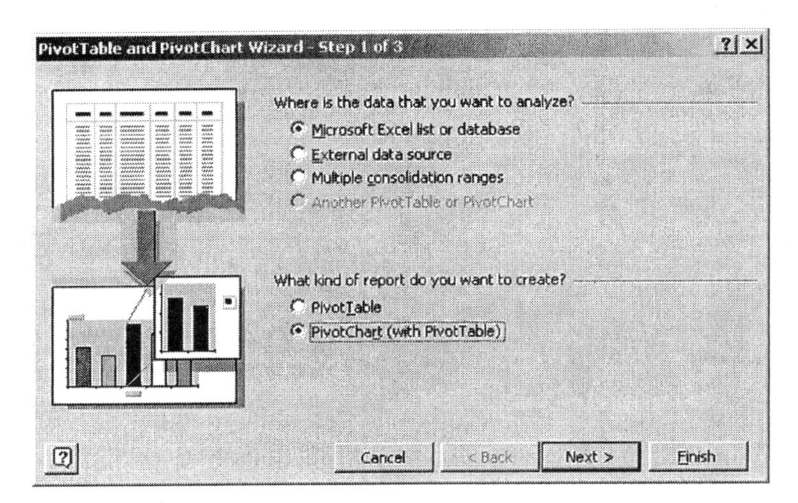

4. The data range is automatically placed in the Range window of the dialog box. It should read $A$1:$A$41. Make any necessary revisions to the range, and then click **Next>**.

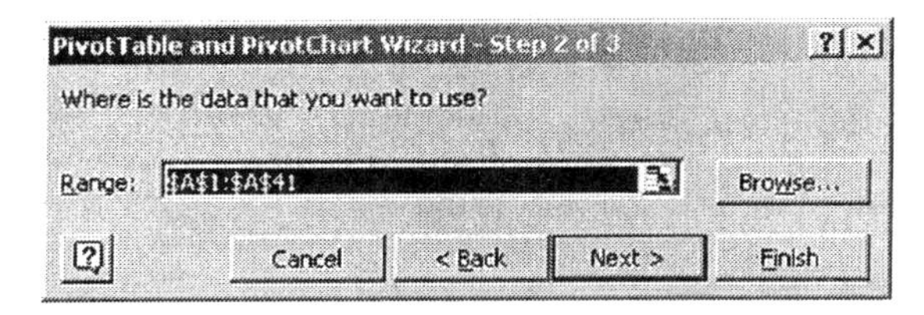

5. You can place the table and chart in a new worksheet or in the existing worksheet. For this example, select **New worksheet**. Click **Finish**.

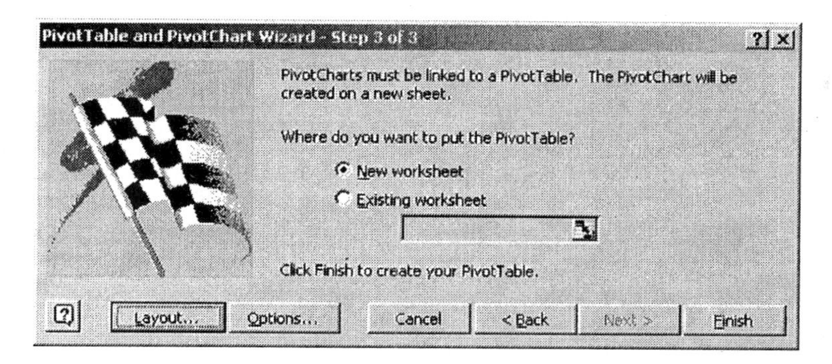

6. The framework for the chart is displayed in a new worksheet. It should look similar to the diagram shown below. Move the cursor to the variable label on the Pivot Table toolbar. For this example, the variable label is C1-T. Click and drag **C1-T** to the area of the diagram labeled "Drop Data Items Here."

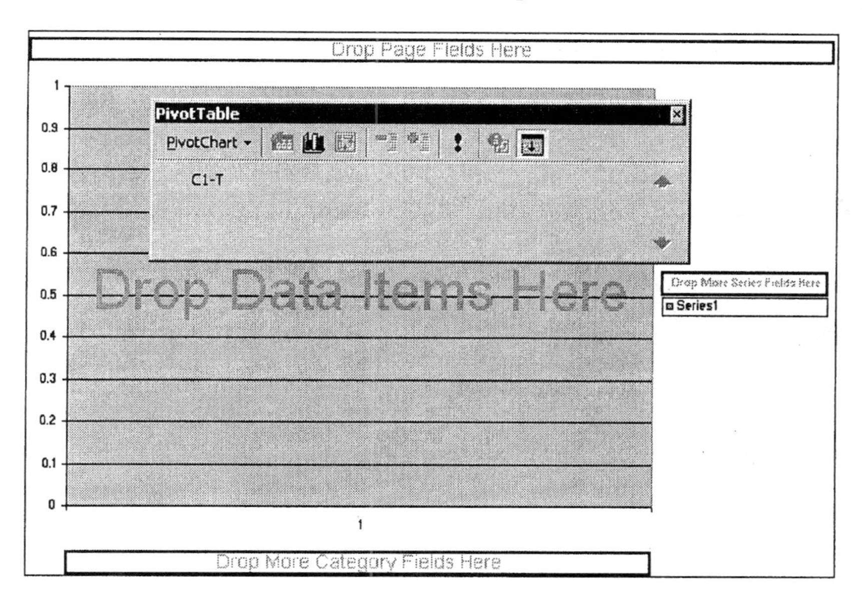

7. Click and drag **C1-T** again, this time to the Category X Axis, right on top of the word "Total" at the bottom of the graph.

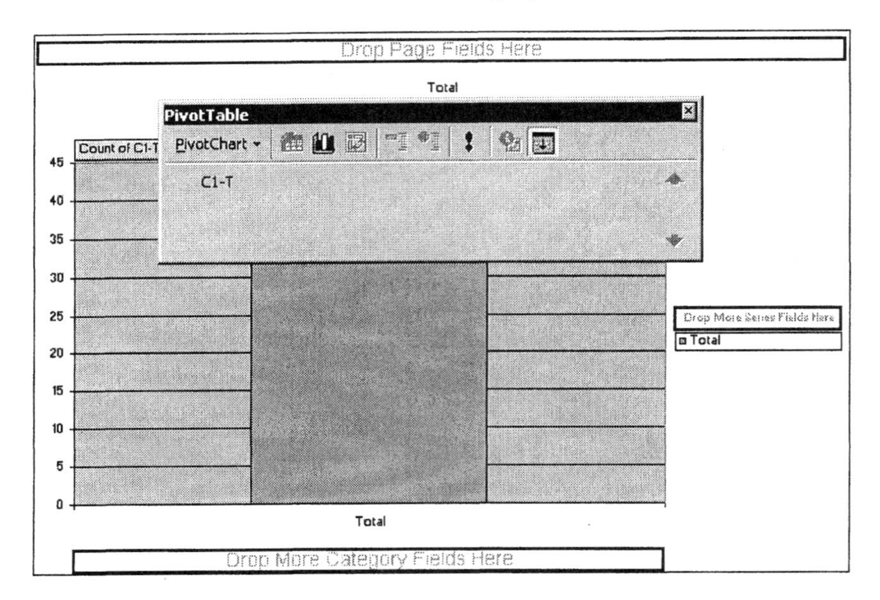

8. Click **Pivot Chart** on the Pivot Table toolbar and then click **Hide Pivot Chart Field Buttons**.

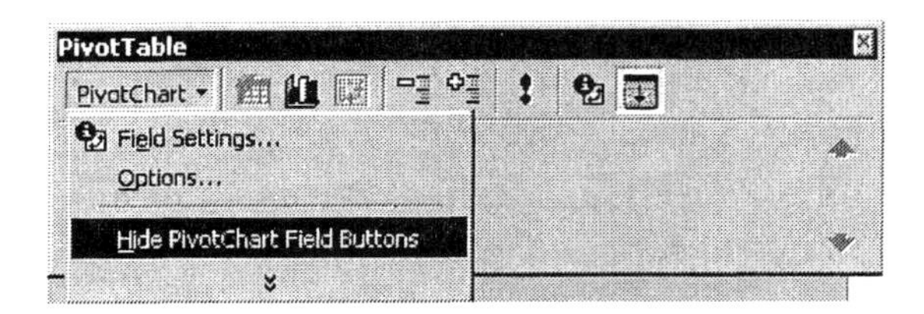

9. Give the chart an appropriate title. Click on the word "Total" at the top of the chart. A frame will appear around the word. Enter the new chart title **2000 Presidential Election Exit Poll in Palm Beach County Florida**. As you are typing, the words will appear in the formula bar near the top of the screen. When you are finished, press **[Enter]**.

10. Next you will remove the legend. Click on the **Total** legend entry to the right of the graph. Black handles will appear around the edges of the legend. Press [**Delete**].

11. Give the vertical axis an appropriate label. **Right click** in the gray area of the graph and select **Chart Options** from the shortcut menu that appears.

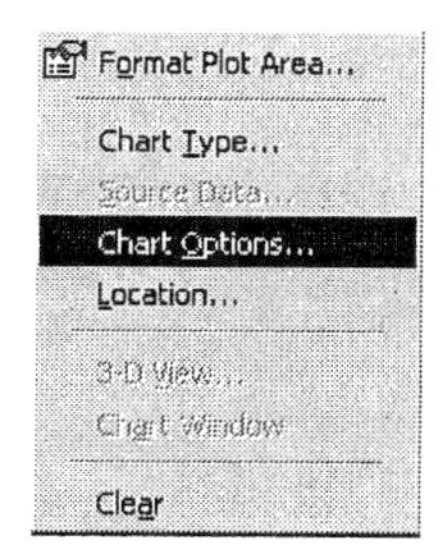

12. Click the **Titles** tab at the top of the Chart Options dialog box. In the Value (Y) axis window, enter **Frequency**. Click **OK**. Your completed bar graph should look similar to the one shown below.

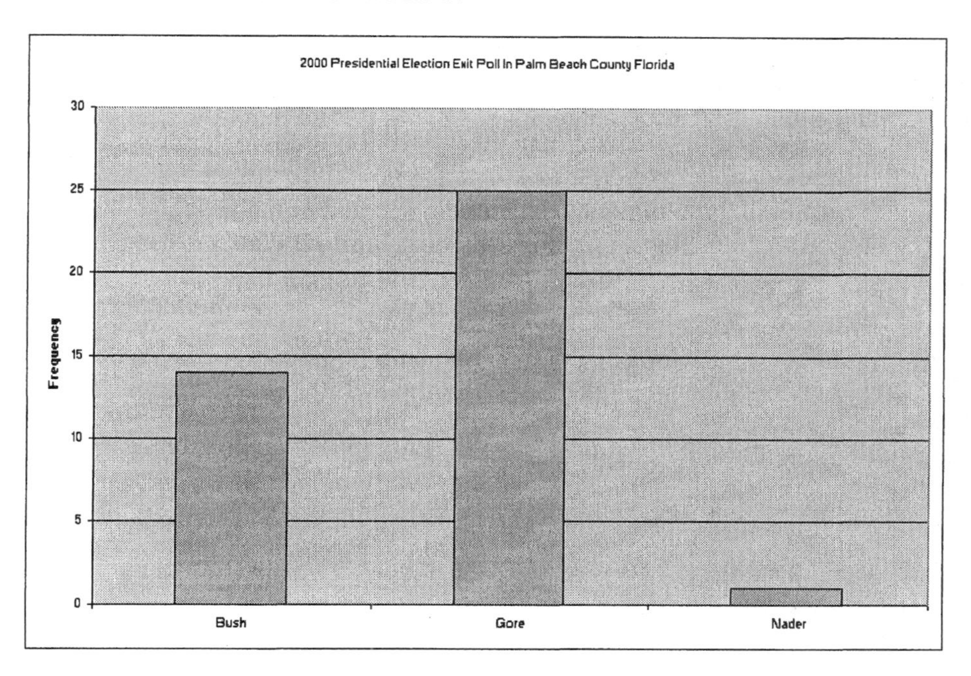

13. The frequency distribution has already been constructed and placed in a new worksheet. Click on the **Sheet1** tab near the bottom of the screen. Your frequency distribution should look similar to the one displayed below.

	A	B	C	D	E	F	G
1							
2							
3	Count of C1-T						
4	C1-T	Total					
5	Bush	14					
6	Gore	25					
7	Nader	1					
8	Grand Total	40					

14. You will use this frequency distribution to construct a relative frequency distribution. First, make a copy of the entire table on the same worksheet. To do this, move the cursor to the top of cell A3 so that it turns into a black down arrow. Then select **Copy** and click in cell **A11** of the worksheet to place the copy a couple of rows below the original. Click **Paste**.

	A	B	C	D	E	F	G
1							
2							
3	Count of C1-T						
4	C1-T	Total					
5	Bush	14					
6	Gore	25					
7	Nader	1					
8	Grand Total	40					
9							
10							
11	Count of C1-T						
12	C1-T	Total					
13	Bush	14					
14	Gore	25					
15	Nader	1					
16	Grand Total	40					

15. The new table will become the relative frequency distribution. **Right click** in any cell of the new table and select **Field Settings** from the shortcut menu.

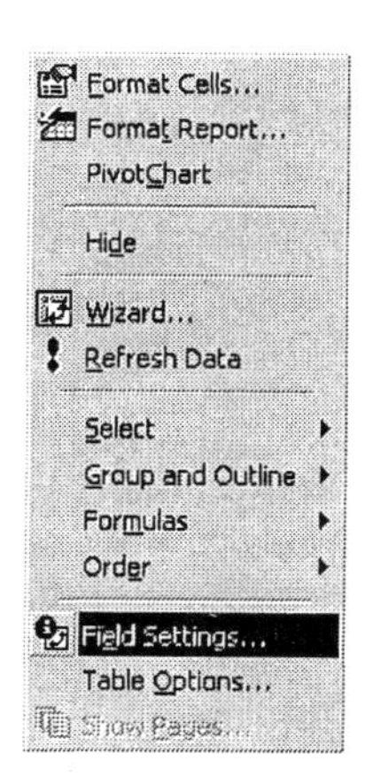

16. Click the **Options** button at the bottom right of the dialog box.

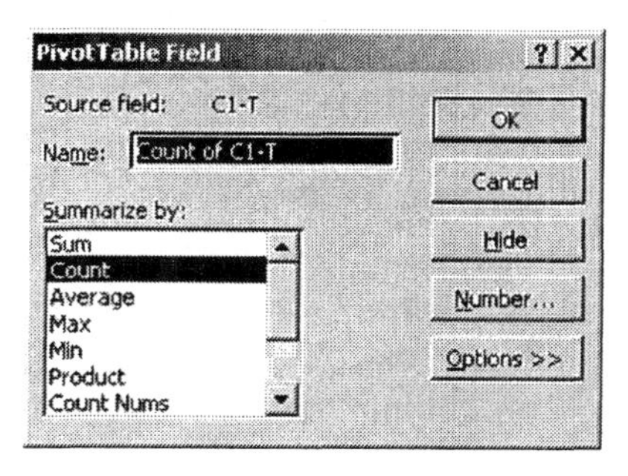

17. At the bottom of the dialog box, in the Show data as window, select **% of column**. Click **OK**.

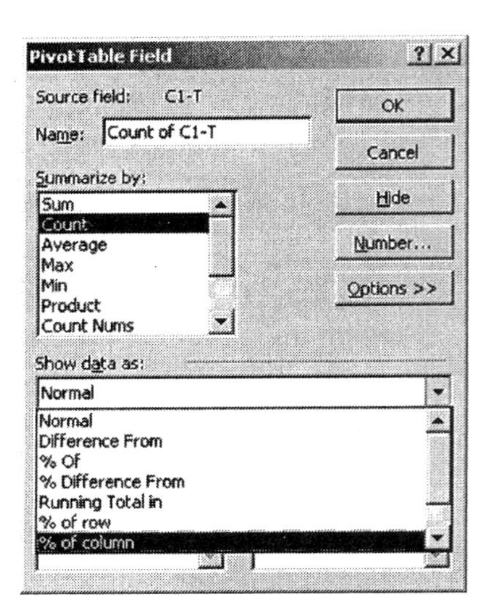

The completed relative frequency distribution with percentages is displayed below.

11	Count of C1-T	
12	C1-T	Total
13	Bush	35.00%
14	Gore	62.50%
15	Nader	2.50%
16	Grand Total	100.00%

18. The final graph you will construct using the 2000 Presidential Election data is a pie chart. You will use the relative frequency distribution to create the pie chart. To begin, **right click** in any cell in the relative frequency distribution table. Select **Pivot Chart** from the shortcut menu.

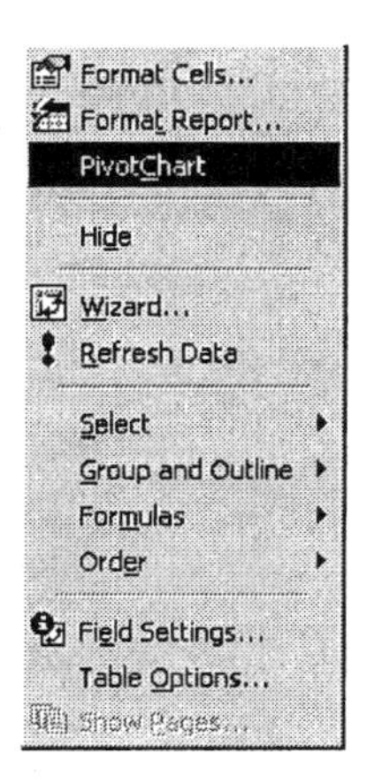

19. A default bar graph of the data will appear. To change it to a pie chart, **right click** in the gray area of the graph and select **Chart Type** from the shortcut menu.

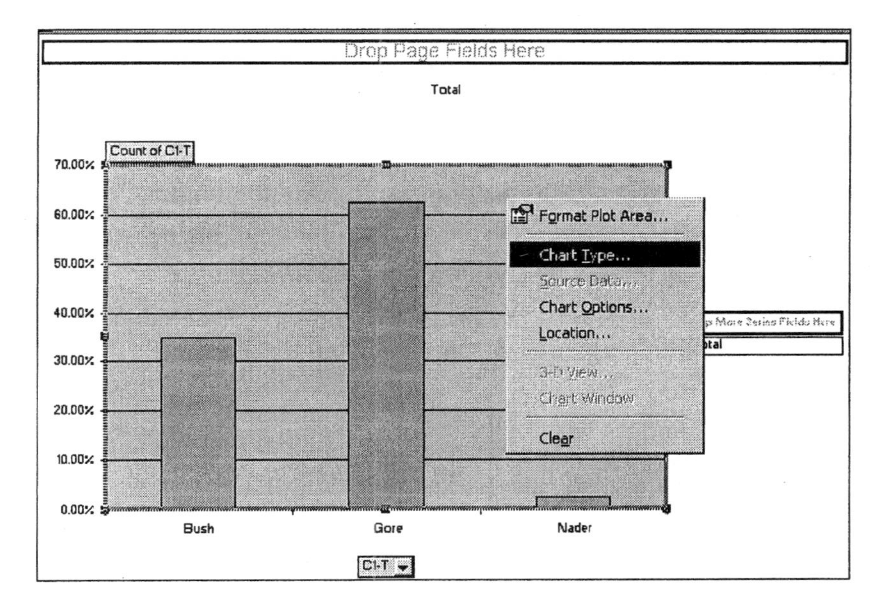

20. Under Chart type at the left of the dialog box, click **Pie** to select it. Then, at the right of the dialog box, under Chart sub-type, click on the first diagram at the far left of the top row to select it. Click **OK**.

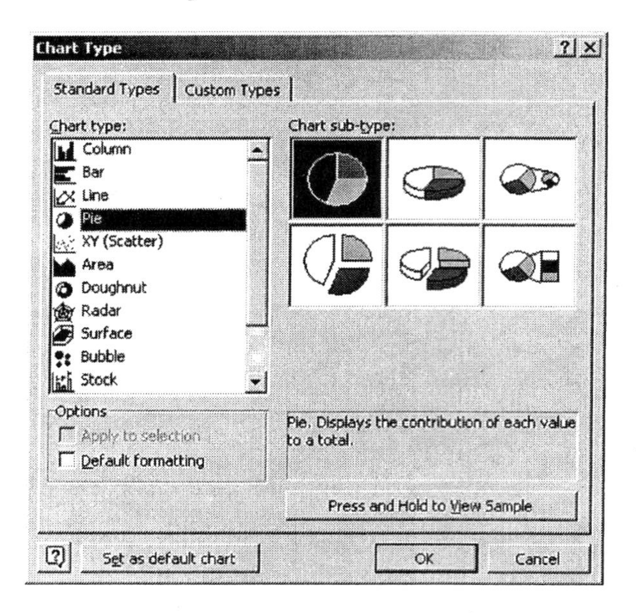

21. To add an appropriate title and make other modifications, **right click** in the white area surrounding the chart and select **Chart Options** from the shortcut menu.

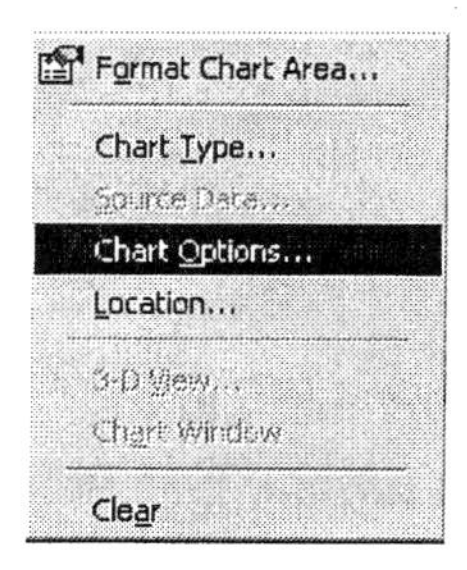

22. Click the **Titles** tab at the top of the dialog box. For the Chart title, enter **2000 Presidential Election Exit Poll in Palm Beach County Florida**.

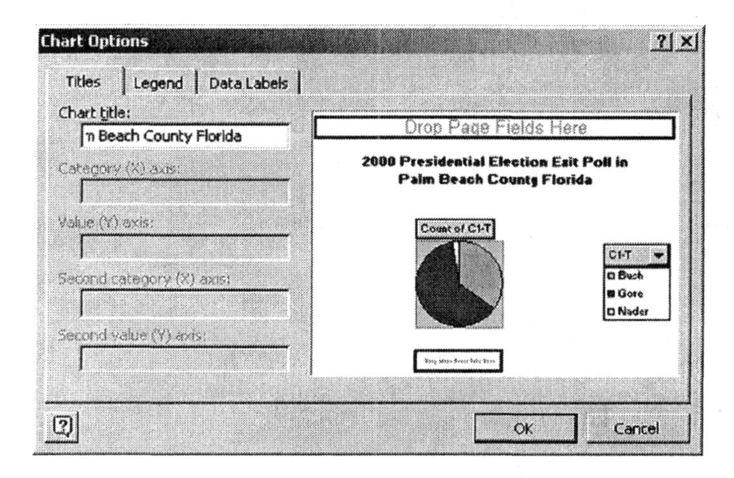

23. Click the **Legend** tab at the top of the dialog box. Click in the box to the left of **Show legend** to remove the checkmark.

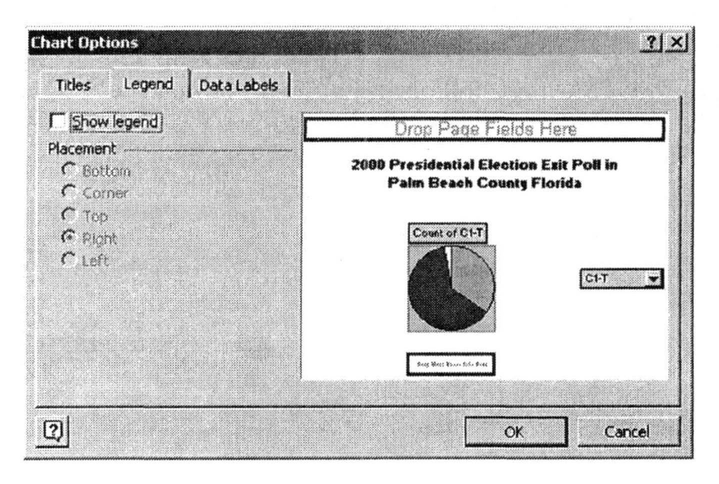

24. Click the **Data Labels** tab at the top of the dialog box. Under Data labels, click the button to the left of **Show label and percent** to select it. At the bottom left of the dialog box, click in the box to the left of **Legend key next to label**. Also click in the box to the left of **Show leader lines**. Click **OK**.

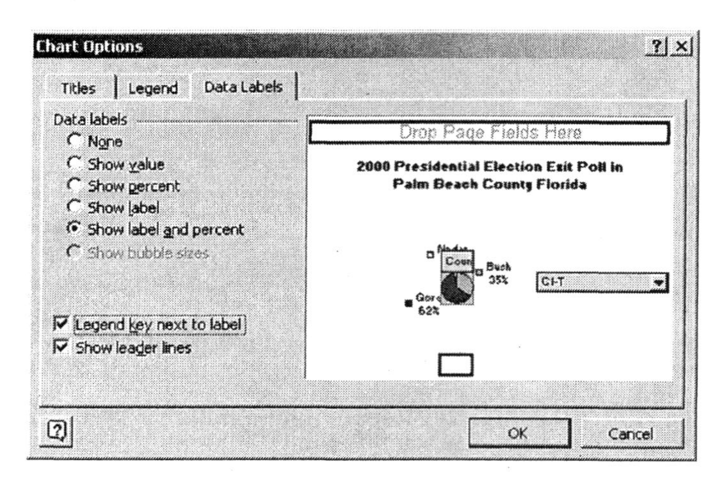

25. As a final step, hide the field buttons. To do this, on the chart, **right click** on the **Count of C1-T** field button. Select **Hide Pivot Chart Field Buttons** from the shortcut menu.

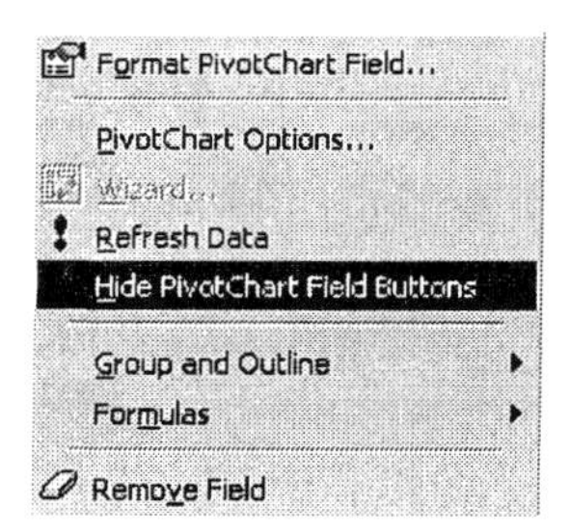

Your completed pie chart should look similar to the one shown below.

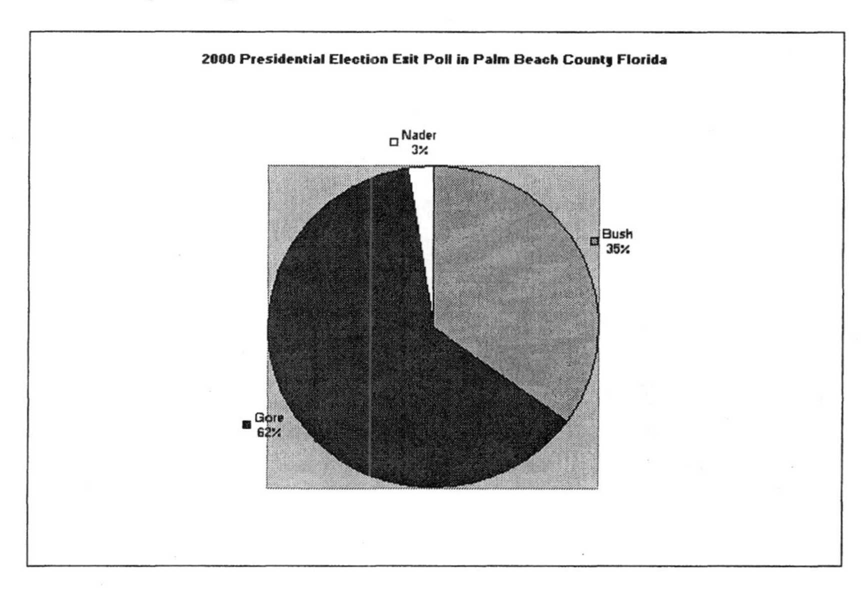

◄

Section 2.2 Organizing Quantitative Data I

► Exercise 17 (pg. 80)

Constructing a Frequency Histogram

You will use the Wait Time data to learn how to construct a frequency histogram.

1. Open worksheet "2_2_17" in the Chapter 2 folder. The first few lines are shown below.

	A	B	C	D	E	F	G
1	Wait Time						
2	11						
3	4						
4	13						

2. Sort the data in ascending order. In the sorted data set, you can see that the minimum number of customers is three. Scroll down to find the maximum number of customers. The maximum number, displayed in cell A41, is 14.

For instructions on how to sort, refer to Sections GS 5.1 and GS 5.2.

	A	B	C	D	E	F	G
1	Wait Time						
2	3						

3. Enter the label **Lower Limit** in cell B1 of the worksheet. You will use the minimum value as the lower limit of the first class, so enter **3** in cell B2. You will use a class width of 2. Calculate the remaining lower limits by adding the class width of 2 to the lower limit of each previous class. You will use a formula to do these computations in the Excel worksheet. Click in cell **B3** and key in **=B2+2** as shown below. Press [**Enter**].

	A	B	C	D	E	F	G
1	Wait Time	Lower Limit					
2	3	3					
3	3	=B2+2					

4. Click in cell **B3** (where 5 now appears) and copy the contents of cell B3 to cells B4 through B8. Because the maximum number of waiting customers is 14, you have calculated one more lower limit than is needed for the histogram. The value of 15, however, will be used when calculating the upper limit of the last class.

	A	B	C	D	E	F	G
1	Wait Time	Lower Limit					
2	3	3					
3	3	5					
4	4	7					
5	4	9					
6	4	11					
7	5	13					
8	5	15					

5. Enter the label **Upper Limit** in cell C1. The upper limit is equal to one less than the lower limit of the next higher class. To do these calculations, you will enter a formula in the Excel worksheet. Click in cell **C2** and enter the formula **=B3-1** as shown in the worksheet below. Press [**Enter**].

	A	B	C	D	E	F	G
1	Wait Time	Lower Limi	Upper Limit				
2	3	3	=B3-1				

6. Copy the formula in C2 (where 4 now appears) to cells C3 through C7. You will use these upper limits for the bins when you construct the histogram chart.

	A	B	C	D	E	F	G
1	Wait Time	Lower Limi	Upper Limit				
2	3	3	4				
3	3	5	6				
4	4	7	8				
5	4	9	10				
6	4	11	12				
7	5	13	14				

7. At the top of the screen, click **Tools** and select **Data Analysis**. In the Data Analysis dialog box, select **Histogram** and click **OK**.

If Data Analysis does not appear as a choice in the Tools menu, you will need to load the Microsoft Excel Analysis ToolPak add-in. Follow the procedure in Section GS 8.1 before continuing.

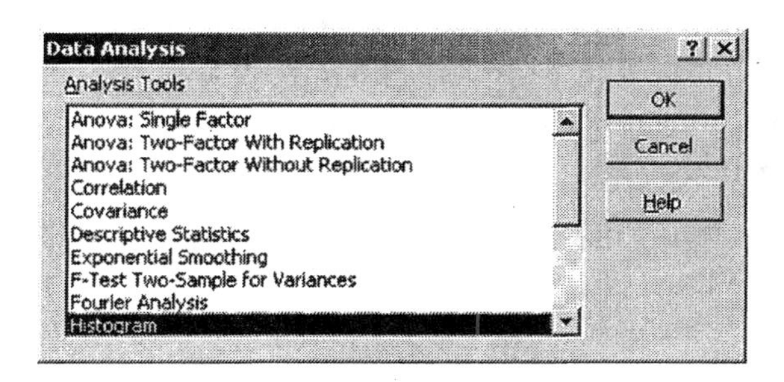

8. Complete the fields in the histogram dialog box as shown below. Be sure to select **Labels** and **Chart Output**. Click **OK**.

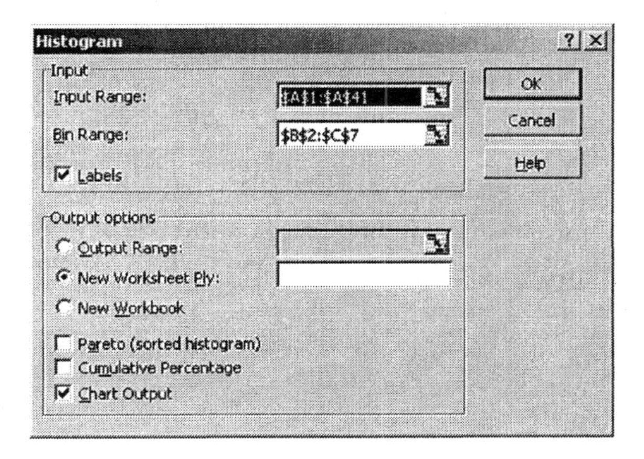

9. The histogram is placed in a new worksheet. You will now follow steps to modify this histogram so that it is presented in a more informative manner. Begin by making the chart taller so that it is easier to read. To do this, first click within the figure near a border. Black square handles appear. Click on the center handle at the bottom border of the figure and drag it down a few rows.

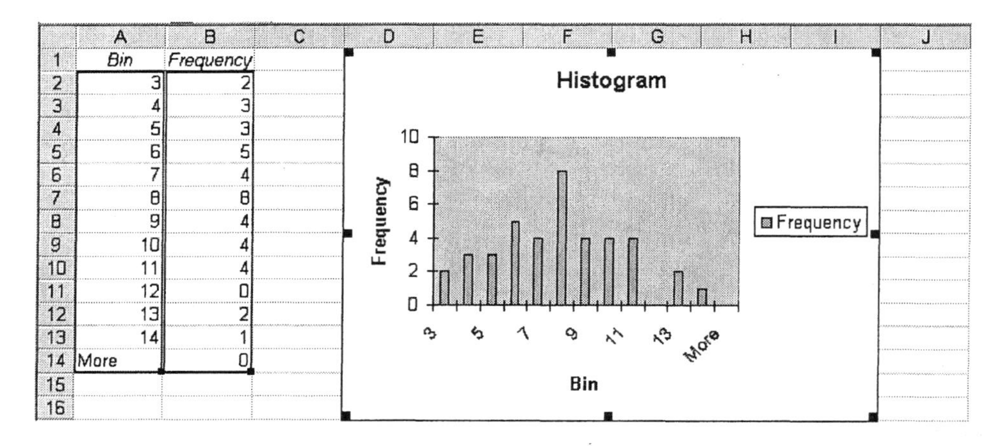

10. Remove the “More” category from the X axis. To do this, **right click** in the gray plot area of the histogram and select **Source Data** from the shortcut menu that appears.

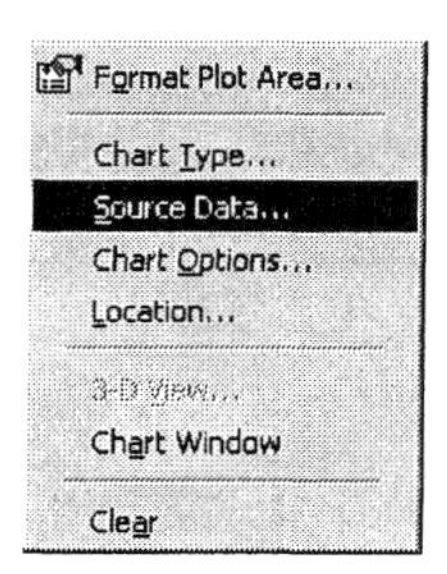

11. Click the Series tab at the top of the Source Data dialog box. “More” appears in cell A14 of the worksheet and its zero frequency appears in cell B14. To exclude that information, edit the entry in the Values window and the entry in the Category (X) axis labels window. The entry in the Values window should read **=Sheet1!$B$2:$B$13**. The entry in the Category (X) axis labels window should read **=Sheet1!$A$2:$A$13**. Click **OK**.

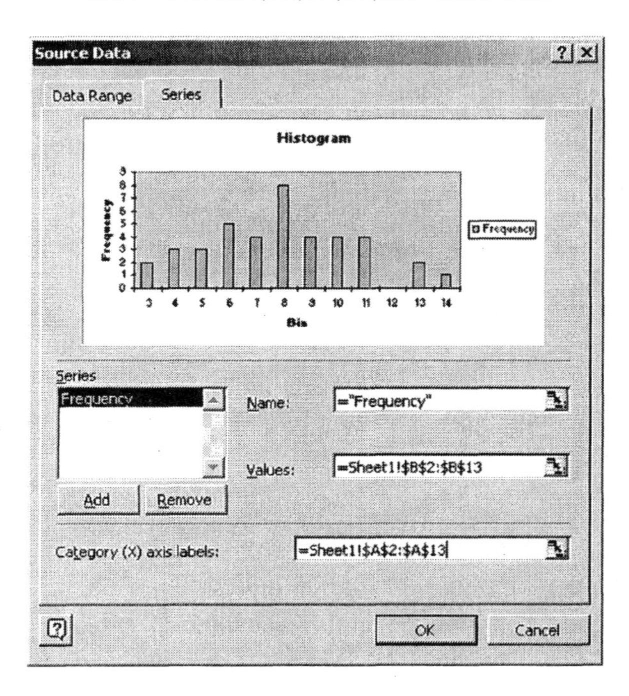

12. **Right click** in the gray plot area of the histogram and select **Chart Options** from the menu that appears.

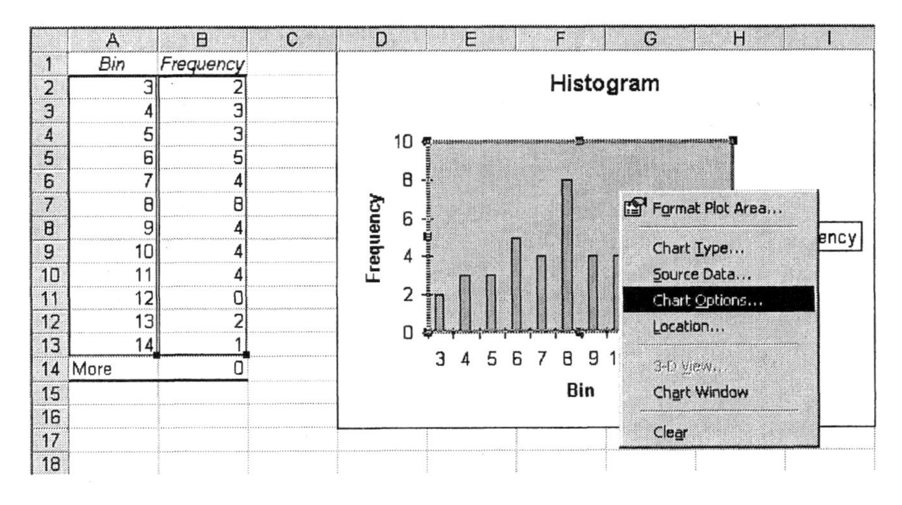

13. Click the **Titles** tab at the top of the Chart Options dialog box. In the Chart title field, replace "Histogram" with **Customers Waiting for a Table**. In the Category (X) axis field, enter **Number of Customers**.

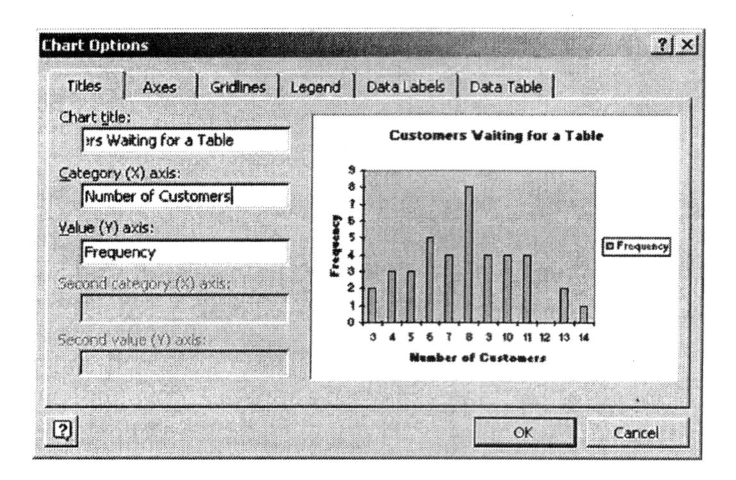

14. Click the **Gridlines** tab at the top of the dialog box. Under Value (Y) axis, click in the box next to **Major gridlines** so that a checkmark appears there.

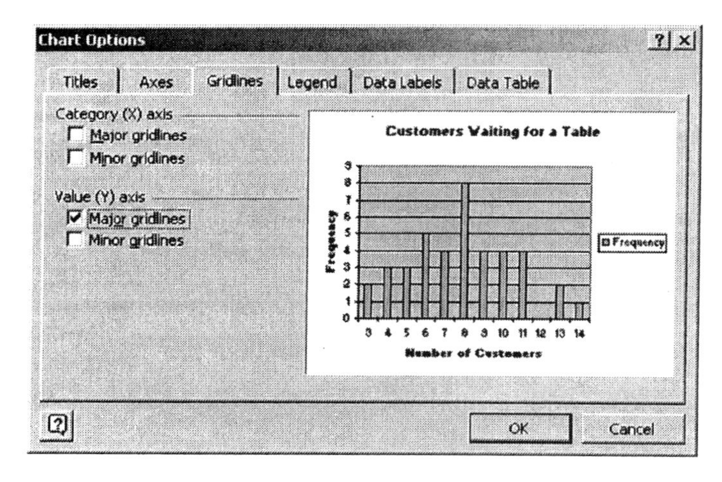

15. Click the **Legend** tab at the top of the dialog box. Click in the box to the left of **Show legend** to remove the checkmark. The removal of the checkmark will delete the frequency legend from the right side of the histogram chart. Click **OK**.

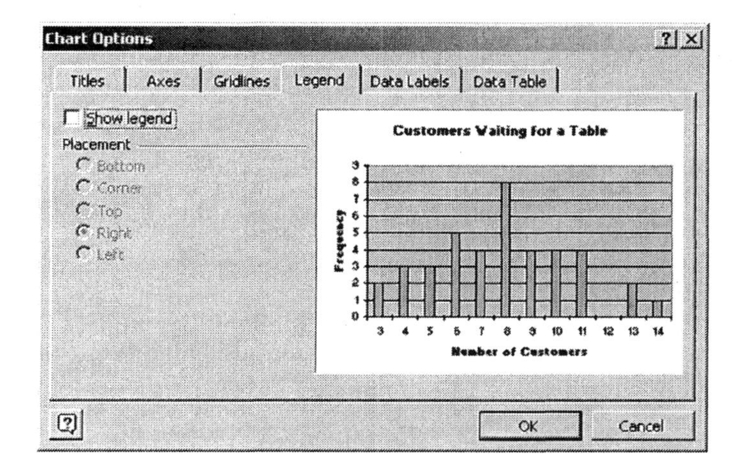

16. Remove the space between the vertical bars. **Right click** directly on one of the vertical bars. Select **Format Data Series** from the shortcut menu that appears.

17. Click the **Options** tab at the top of the Format Data Series dialog box. Change the value in the Gap width box to 0. Click **OK**.

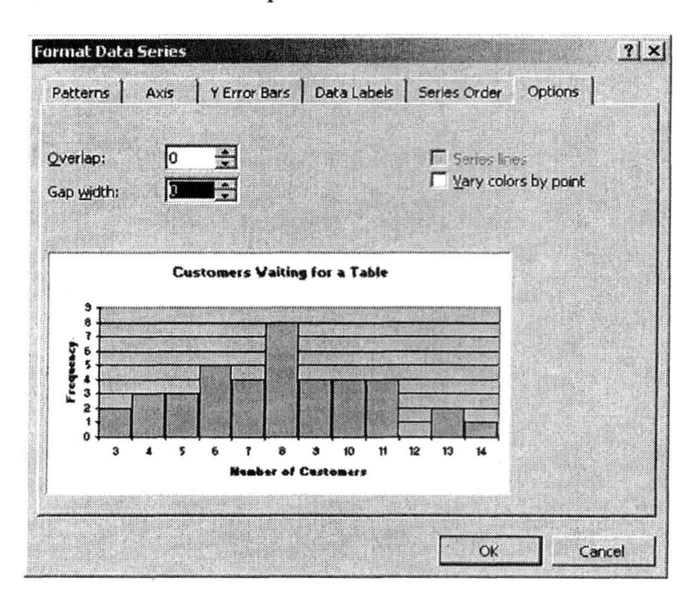

Your completed histogram should look similar to the one displayed below.

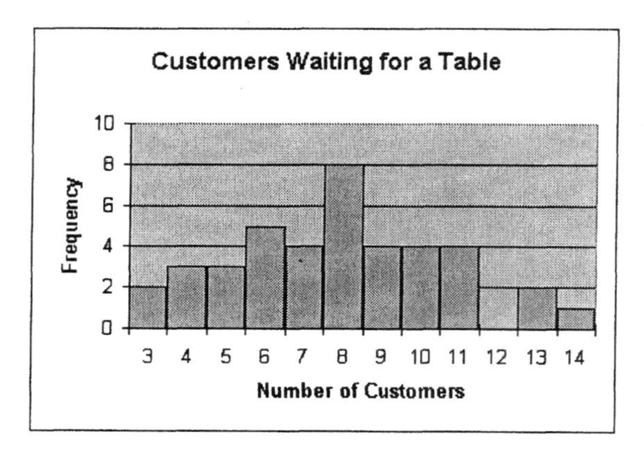

◀

► Exercise 25 (pg. 82) Constructing a Stem-and-Leaf Diagram

If the PHStat add-in has not been loaded, you will need to load it before continuing. Follow the instructions in Section GS 8.2.

1. Open worksheet "2_2_25" in the Chapter 2 folder. The first few rows are shown below.

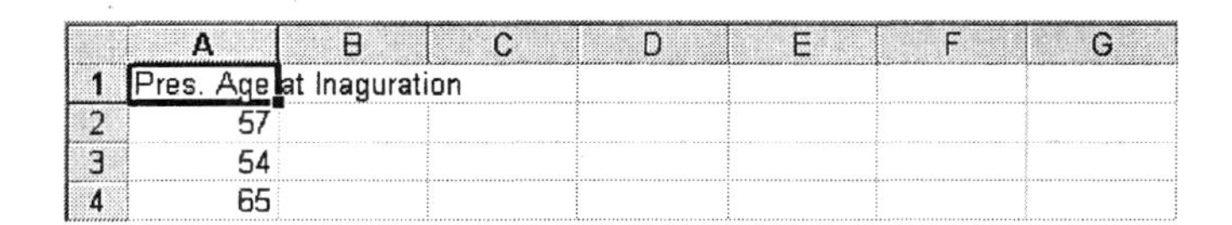

	A	B	C	D	E	F	G
1	Pres. Age at Inaguration						
2	57						
3	54						
4	65						

2. At the top of the screen, click **PHStat**. Select **Descriptive Statistics →Stem-and-Leaf Display**.

3. In the Stem-and-Leaf Display dialog box, click in the **Variable Cell Range Window** so that the flashing I-beam appears there. Then click and drag over the Age at Inauguration data (cell A1 through cell A44) to enter the range in the Variable Cell Range window. Click **OK**.

*If you prefer to key in the range rather than clicking and dragging, you can key in **A1:A44**.*

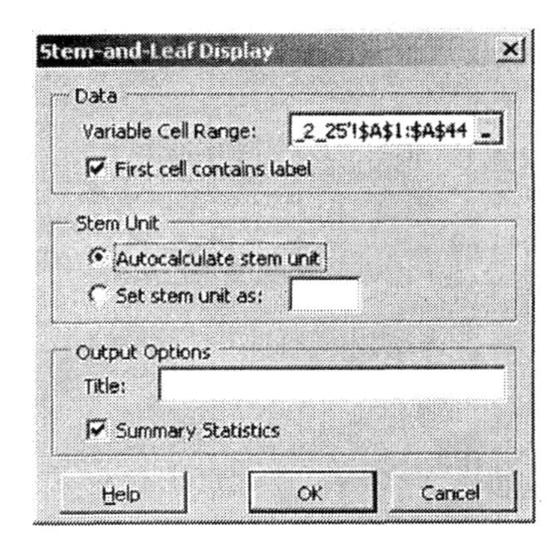

Your completed stem-and-leaf display output should look similar to the display shown below.

	A	B	C	D	E	F	G	H	I
1				Stem-and-Leaf Display					
2				for Pres. Age at Inauguration					
3				Stem unit: 10					
4									
5	Statistics								
6	Sample Size	43		4	2 3 6 6 7 8 9 9				
7	Mean	54.81395		5	0 0 1 1 1 1 2 2 4 4 4 4 4 5 5 5 5 6 6 6 7 7 7 7 8				
8	Median	55		6	0 1 1 1 2 4 4 5 8 9				
9	Std. Deviation	6.234527							
10	Minimum	42							
11	Maximum	69							

Section 2.3 Organizing Quantitative Data II

► Exercise 5 (pg. 90) Constructing a Histogram, Frequency Polygon, Relative Frequency Polygon, and Cumulative Percentage Polygon (Ogive)

If the PHStat add-in has not been loaded, you will need to load it before continuing. Follow the instructions in Section GS 8.2.

1. Open worksheet "2_2_T12" in the Chapter 2 folder. The first few rows are shown below.

	A	B	C	D	E	F	G
1	3-Yr Mutual Fund Rate of Return						
2	27.4						
3	12.7						
4	22.6						

2. Sort the data in ascending order so that you can identify the minimum and maximum values. In the sorted data set, you can see that the minimum three-year-rate is 10.8. Scroll down to cell A41 to find the maximum. The maximum is 47.7.

For instructions on how to sort, refer to Sections GS 5.1 and GS 5.2.

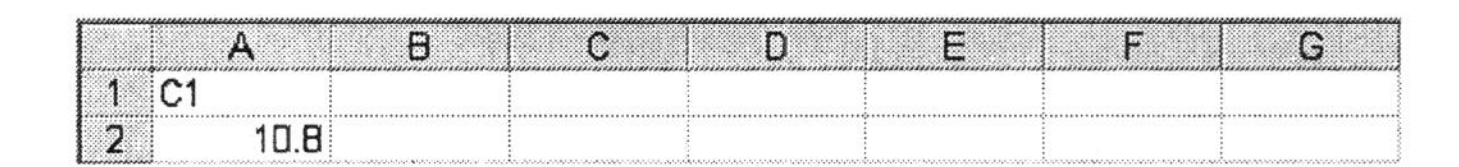

	A	B	C	D	E	F	G
1	C1						
2	10.8						

3. This procedure requires you to enter the class maximum values in a column of the worksheet. You will use a class width of 5 and maximum values 14.9, 19.9, 24.9,…, 49.9. To be consistent with the dialog box, you will name these values Bins. Enter **Bins** in cell B1 of the worksheet. Enter **14.9** in cell B2.

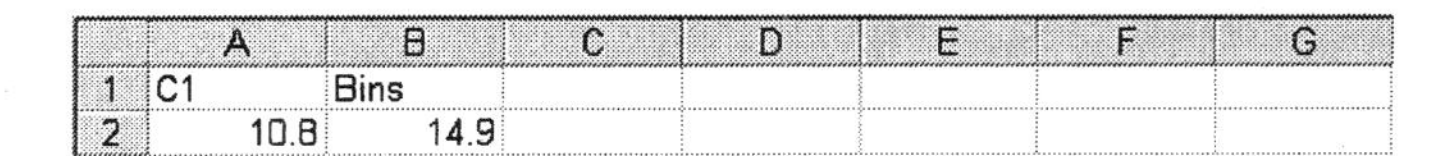

	A	B	C	D	E	F	G
1	C1	Bins					
2	10.8	14.9					

4. Click in cell **B2** where 14.9 now appears. At the top of the screen, click **Edit**. Select **Fill** → **Series**.

5. Complete the Series dialog box as shown below. Be sure to select **Columns** under the Series in choices. The step value is **5**, and you want to stop at **49.9**. Click **OK**.

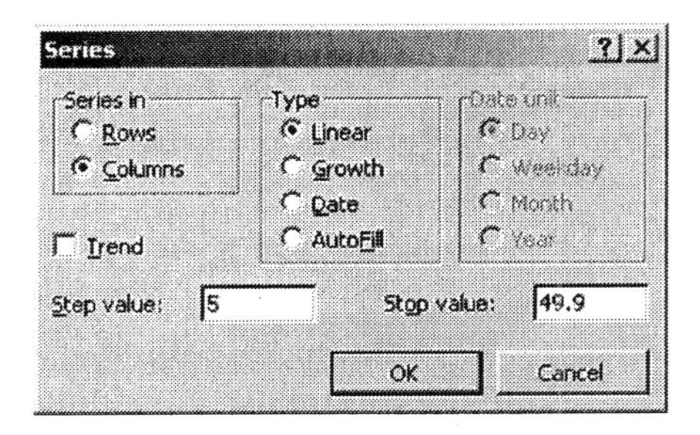

6. This procedure requires you to enter the midpoints. The first class is 10 to 14.9. To find the midpoint, first subtract 10 from 14.9 and divide by 2. This is equal to 2.45. The midpoint of the first class is equal to 10 plus 2.45 or 12.45. The remaining midpoints proceed in increments of 5 up to 47.45. To be consistent with the dialog box, name these values Midpoints. Enter **Midpoints** in cell C1. Enter **12.45** in cell C2.

	A	B	C	D	E	F	G
1	C1	Bins	Midpoints				
2	10.8	14.9	12.45				

7. Click in cell **C2** where 12.45 now appears. At the top of the screen, click **Edit**. Select **Fill → Series.**

8. Complete the Series dialog box as shown below. Be sure to select **Columns**. The step value is **5**, and you want to stop at **47.45**. Click **OK**.

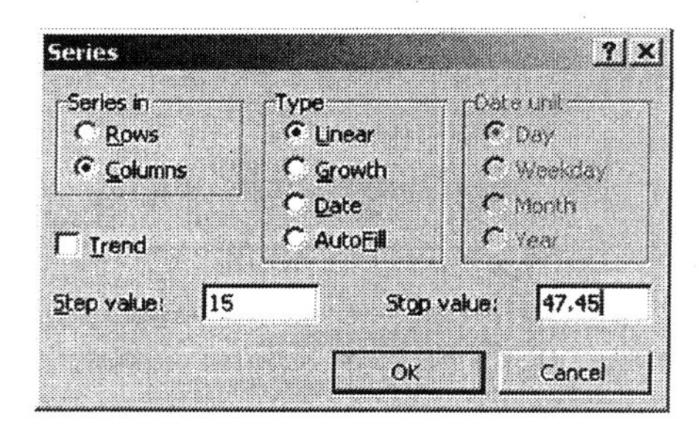

9. Your worksheet should now look like the line displayed below. At the top of the screen, click **PHStat**. Select **Descriptive Statistics → Histograms & Polygons**.

	A	B	C	D	E	F	G
1	C1	Bins	Midpoints				
2	10.8	14.9	12.45				
3	10.9	19.9	17.45				
4	11.6	24.9	22.45				
5	12.7	29.9	27.45				
6	12.8	34.9	32.45				
7	14.7	39.9	37.45				
8	14.7	44.9	42.45				
9	15.8	49.9	47.45				

10. Complete the dialog box as shown below. Click and drag over the variable, bins, and midpoint values in the worksheet to enter them in the appropriate windows in the dialog box. Select **Single Group Variable**. Under Output Options, select all four: **Histogram**, **Frequency Polygon**, **Percentage Polygon**, and **Cumulative Percentage Polygon (Ogive)**. Click **OK**.

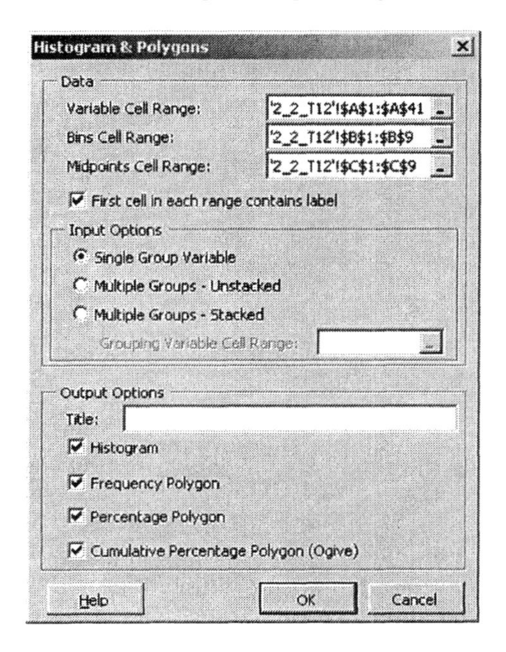

11. The four requested graphs are generated and placed in four different worksheets. The sheet labeled Polygon 1 contains the frequency polygon.

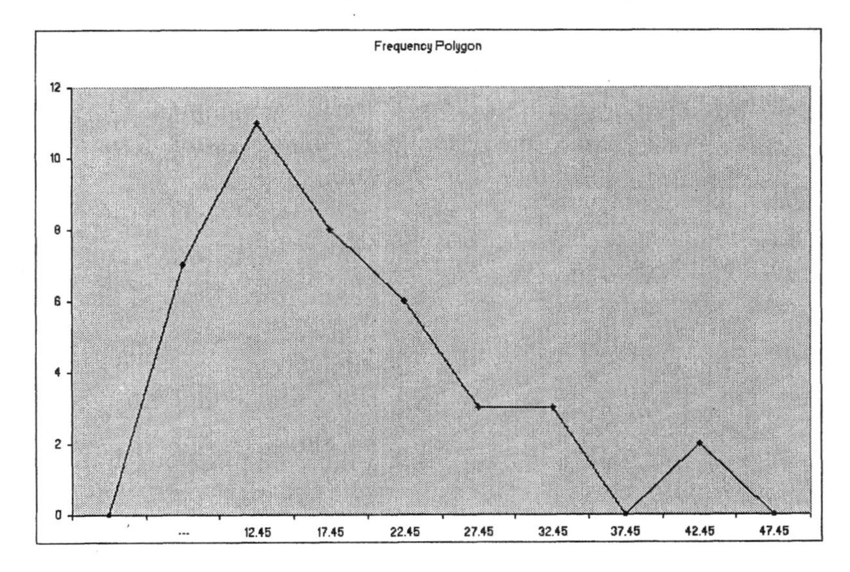

The sheet labeled Polygon 2 contains the percentage polygon.

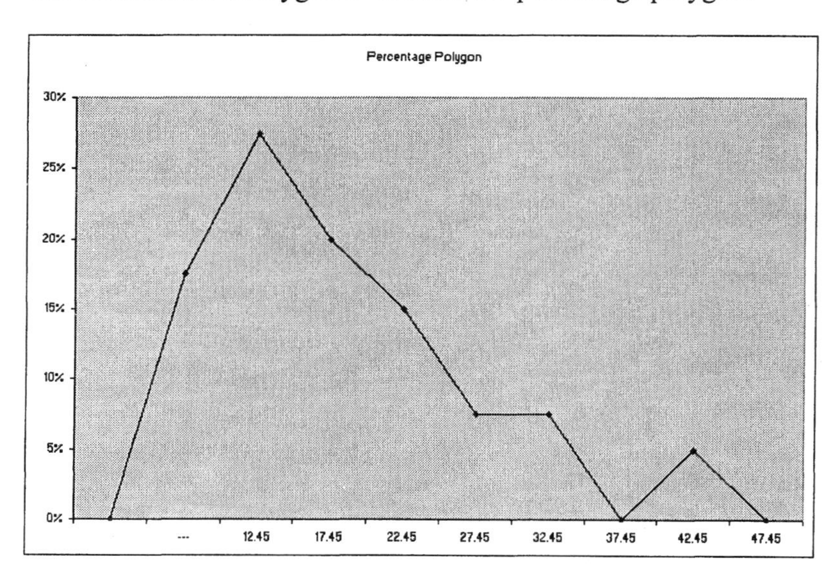

The sheet labeled Polygon 3 contains the cumulative percentage polygon (ogive).

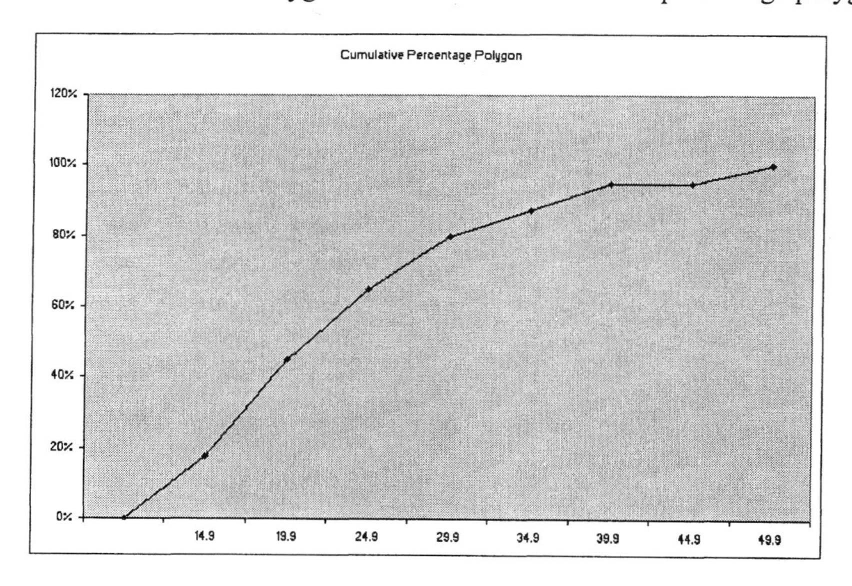

The sheet labeled Frequency contains the histogram and summary data.

	A	B	C	D	E
1	Frequencies (C1)				
2	*Bins*	*Frequency*	*Percentage*	*Cumulative %*	*Midpts*
3		0	0	0	
4	14.9	7	17.50%	17.50%	---
5	19.9	11	27.50%	45.00%	12.45
6	24.9	8	20.00%	65.00%	17.45
7	29.9	6	15.00%	80.00%	22.45
8	34.9	3	7.50%	87.50%	27.45
9	39.9	3	7.50%	95.00%	32.45
10	44.9	0	0.00%	95.00%	37.45
11	49.9	2	5.00%	100.00%	42.45
12		0	0		47.45

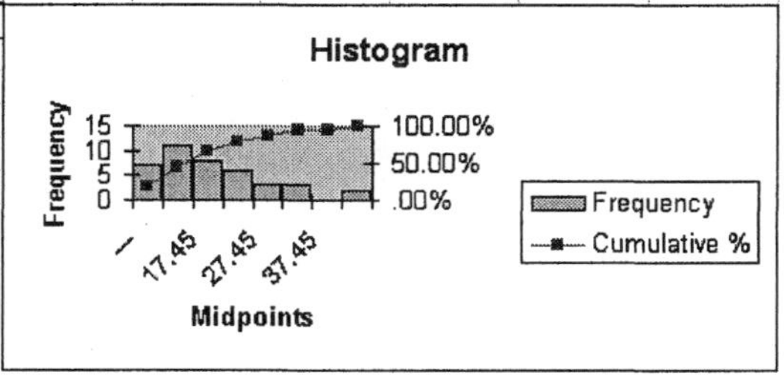

Histogram

Frequency

15
10
5
0

100.00%
50.00%
.00%

17.45
27.45
37.45

Midpoints

Frequency
Cumulative %

Numerically Summarizing Data

CHAPTER 3

Section 3.1 Measures of Central Tendency

► Example 1 (pg. 113) — Computing a Population Mean and a Sample Mean

You will use the home run data to learn how to compute a population mean, generate a random sample, and compute the sample mean.

If the PHStat add-in has not been loaded, you will need to load it before continuing. Follow the instructions in Section GS 8.2.

1. Open worksheet "3_1_Ex1" in the Chapter 3 folder. The first few rows are shown below.

	A	B	C	D	E	F	G
1	Team	Home runs					
2	1. Anahein	158					
3	2. Baltimor	136					
4	3. Boston	198					

2. Click in cell **B16** below the home run data.

3. At the top of the screen, click **Insert** and select **Function**.

4. Under Function category, select **Statistical**. Under Function name, select **AVERAGE**. Click **OK**.

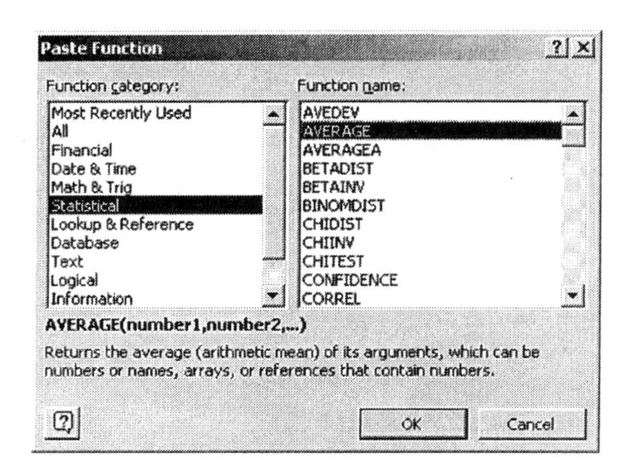

5. You should see the range B2:B15 in the Number 1 window of the dialog box. If this range does not appear in the window, you will need to enter it. Click **OK**.

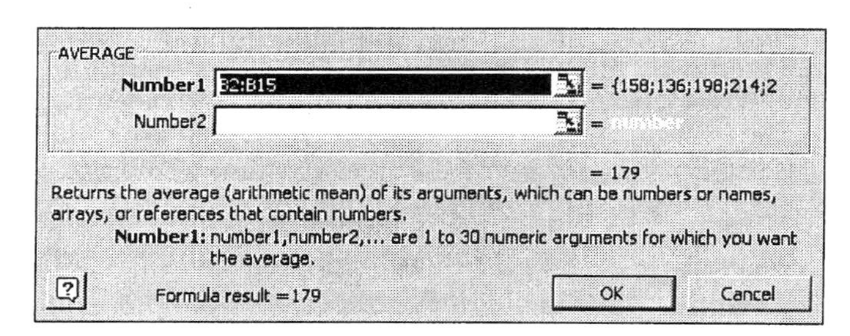

6. You will now see 179 in cell B16. You will want to enter the label **Mean** in cell A16 for reference purposes.

14	13. Texas	246
15	14. Torontc	195
16	Mean	179

7. You will now generate a simple random sample of size n = 5 from this population. At the top of the screen, click **PHStat**. Select **Sampling → Random Sample Generation**.

8. Complete the Random Sample Generator dialog box as shown below. A sample of five values will be selected from the range B1:B15. The first cell of this range contains a label (Home runs). The topmost cell of the output will contain the label "Home Run Sample." Click **OK**.

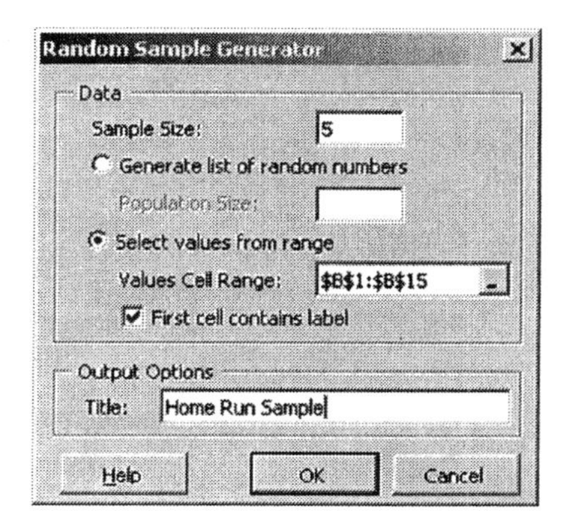

9. The output is displayed in a new worksheet named "RandomNumbers." Because the numbers were generated randomly, it is not likely that your output will be exactly the same. Click in cell **A7** below the sample data.

	A	B	C	D	E	F	G
1	Home Run	Sample					
2	203						
3	121						
4	169						
5	199						
6	164						

10. At the top of the screen, click **Insert** and select **Function**.

11. Under Function category, select **Statistical**. Under Function name, select **AVERAGE**. Click **OK**.

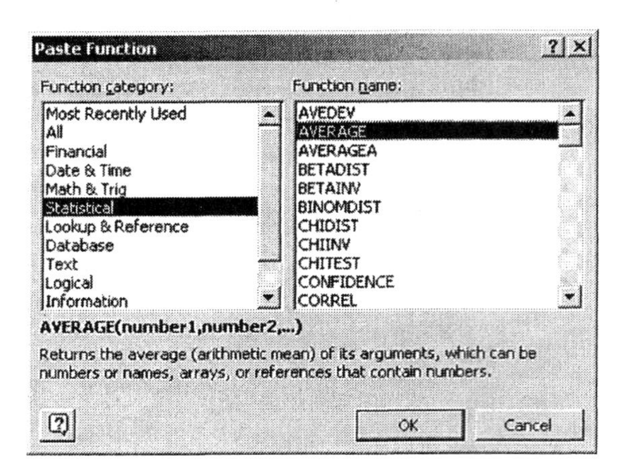

12. You should see the range A1:A6 in the Number 1 window of the dialog box. If this range does not appear, you will need to enter it. Click **OK**.

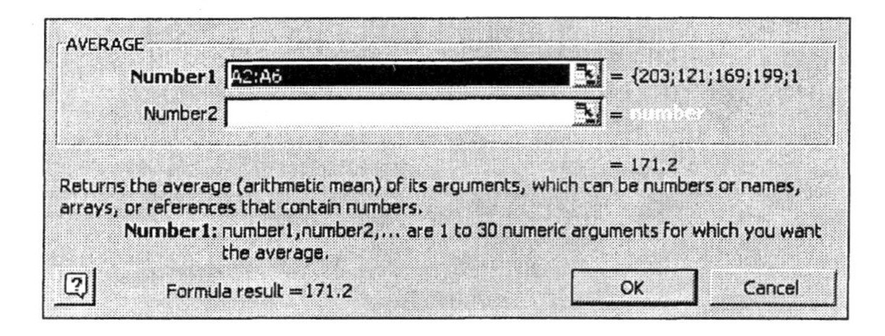

The sample mean will appear in cell A7. The mean for the sample that I generated is 171.2.

	A	B	C	D	E	F	G
1	Home Run Sample						
2	203						
3	121						
4	169						
5	199						
6	164						
7	171.2						

◀

▶ Exercise 5 (pg. 124) — Computing the Mean, Median, and Mode for Various Plot Types

1. Open worksheet "3_1_5" in the Chapter 3 folder. The first few rows are shown below.

	A	B	C	D	E	F	G
1	Sludge Plo	Spring Dis	No Till	Spring Chi	Great Lakes BT		
2	25	32	30	30	28		
3	27	30	26	32	32		
4	33	33	29	26	27		

2. Begin by inserting a column to the left of the data so that you will can label your output. At the top of the screen, click **Insert** and select **Columns.**

3. Enter the labels **Mean**, **Median**, and **Mode** as shown below.

7		27	34	29	29	27	
8	Mean						
9	Median						
10	Mode						

4. Click in cell **B8** where you will place the mean of Sludge Plot.

5. At the top of the screen, click **Insert** and select **Function.**

6. Under Function category, select **Statistical.** Under Function name, select **AVERAGE**. Click **OK**.

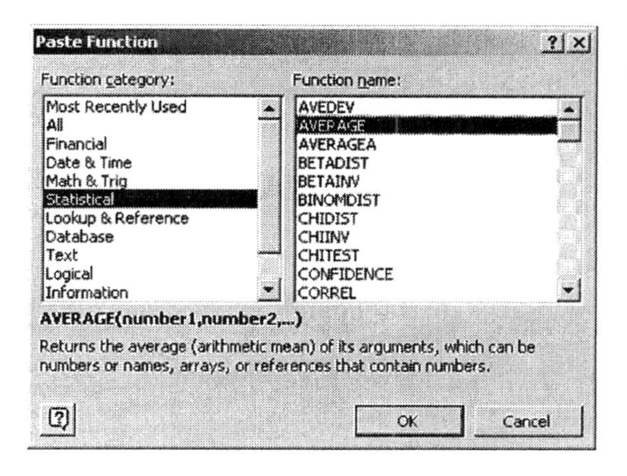

7. You should see the range B2:B7 in the Number 1 window of the dialog box. If this range does not appear, you will need to enter it. Click **OK**.

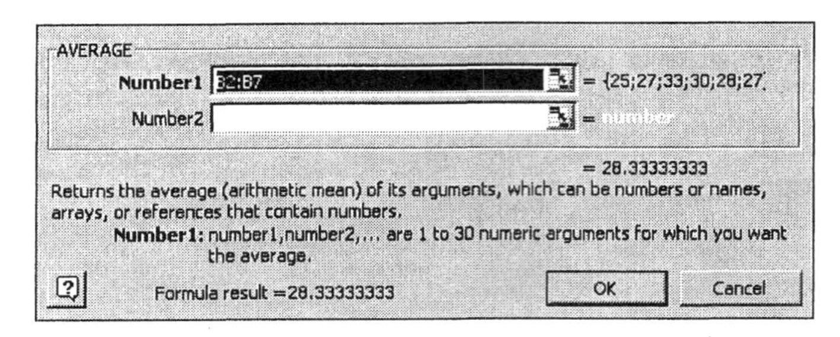

8. You will now see 28.3333 in cell B8. Copy the AVERAGE function in cell B8 to cells C8 through F8. Your output should look similar to the output displayed below.

7		27	34	29	29	27	
8	Mean	28.33333	33	28.5	29.33333	28.83333	
9	Median						
10	Mode						

9. Click in cell **B9** where you will place the Sludge Plot median.

10. At the top of the screen, click **Insert** and select **Function**.

11. Under Function category, select **Statistical**. Under Function name, select **MEDIAN**. Click **OK**.

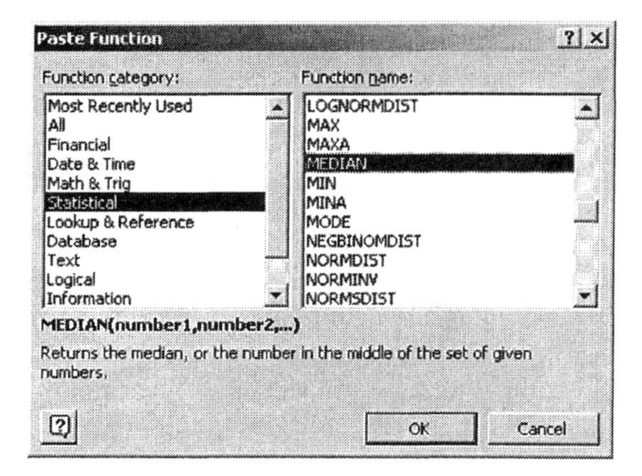

12. The range B2:B8 will automatically appear in the Number 1 window. Change this to **B2:B7** so that the mean in B8 will not be included. Click **OK**.

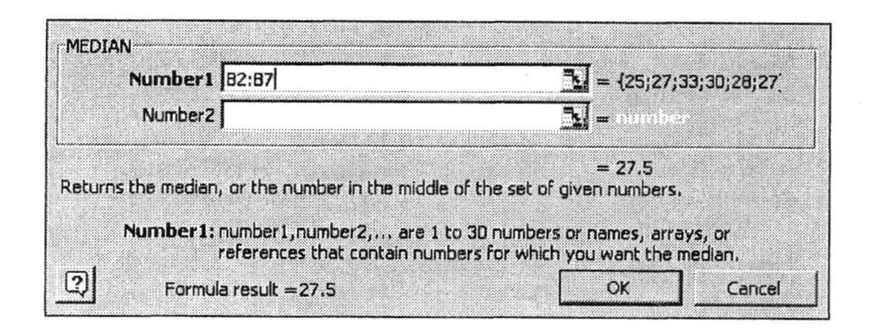

13. Copy the MEDIAN function in B9 to cells C9 through F9.

	A	B	C	D	E	F	G
1		Sludge Plc	Spring Dis	No Till	Spring Chi	Great Lakes BT	
2		25	32	30	30	28	
3		27	30	26	32	32	
4		33	33	29	26	27	
5		30	35	32	28	30	
6		28	34	25	31	29	
7		27	34	29	29	27	
8	Mean	28.33333	33	28.5	29.33333	28.83333	
9	Median	27.5	33.5	29	29.5	28.5	
10	Mode						

14. Click in cell **B10** where you will place the Sludge Plot mode.

15. At the top of the screen, click **Insert** and select **Function**.

16. Under Function category, select **Statistical**. Under Function name, select **MODE**. Click **OK**.

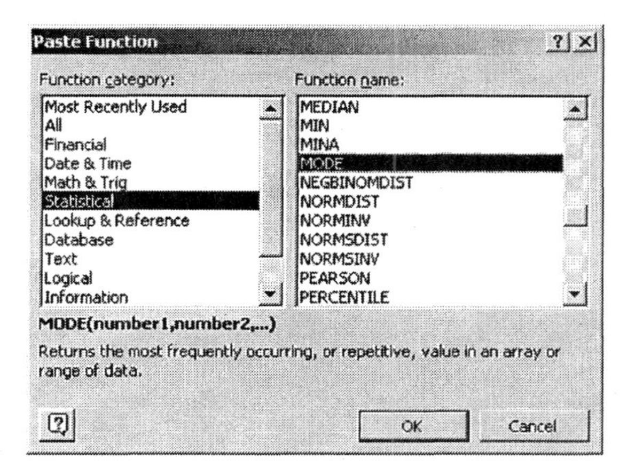

17. The range B2:B9 will automatically appear in the Number 1 window. Change it to B2:B7 so that the mean and median are not included. Click **OK**.

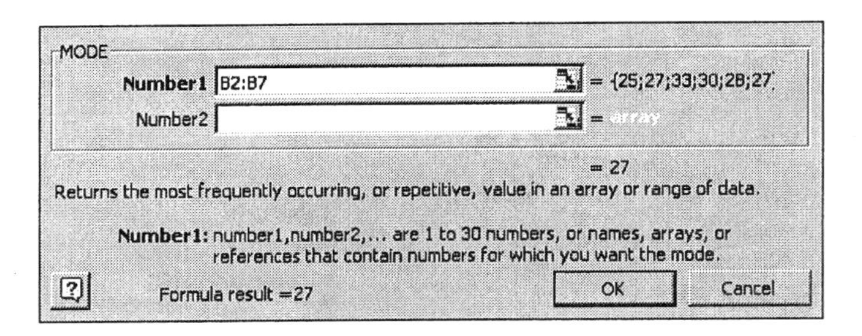

18. Copy the MODE function in B10 to cells C10 through F10. Your completed output should look similar to the output displayed below.

8	Mean	28.33333	33	28.5	29.33333	28.83333	
9	Median	27.5	33.5	29	29.5	28.5	
10	Mode	27	34	29	#N/A	27	

A word of caution is in order regarding the value reported for the mode. Three situations are possible: 1) If all values occur only once in a distribution, Excel will return #N/A. 2) If a variable has only one mode, the MODE function will return that value. 3) If a variable has more than one mode, however, the MODE function will still return only one value. The value that is returned will be the one associated with the modal value that occurs first in the data set. To check for multiple modes, I suggest that you create a frequency distribution.

► Exercise 13 (pg. 125) Computing Mean and Median Waiting Time

1. Open worksheet "3_1_13" in the Chapter 3 folder. The first few rows are shown below.

	A	B	C	D	E	F	G
1	Number of People Waiting						
2	11						
3	4						
4	13						

2. Insert a column to the left of the data where you will place labels for your output. At the top of the screen, click **Insert** and select **Columns**.

3. Enter the labels **Mean** and **Median** in column A as shown below.

39		5
40		8
41		8
42	Mean	
43	Median	

4. Click in cell **B42** where you will place the mean.

5. At the top of the screen, click **Insert** and select **Function**.

6. Under Function category, select **Statistical**. Under Function name, select **AVERAGE**. Click **OK**.

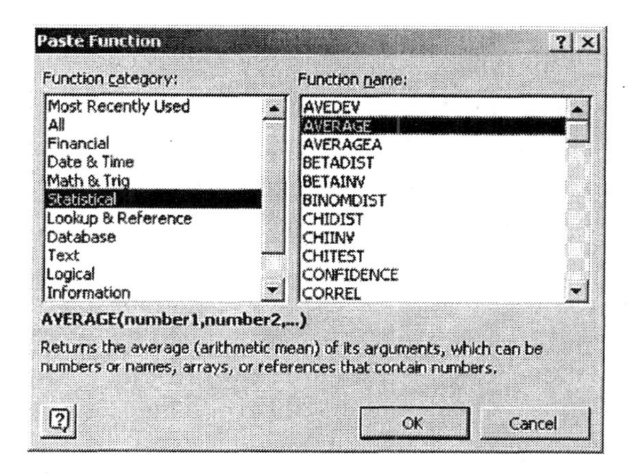

7. You should see the range B2:B41 in the Number 1 window of the dialog box. If this range does not appear, you will need to enter it. Click **OK**.

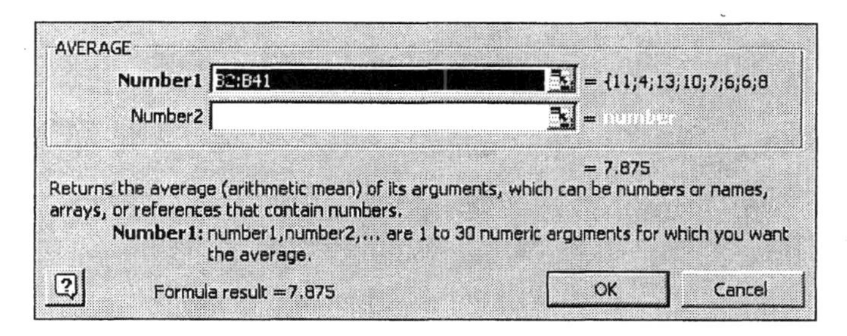

8. You will see 7.875 in cell B42. Next, click in **B43** where you will place the median.

9. At the top of the screen, click **Insert** and select **Function**.

10. Under Function category, select **Statistical**. Under Function name, select **MEDIAN**. Click **OK**.

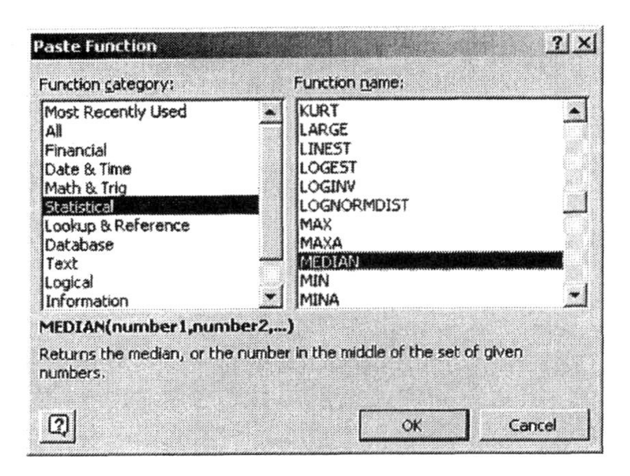

11. The range B2:B42 will automatically appear in the Number 1 window. You will need to change the range to B2:B41 so that the mean is not included. Click **OK**.

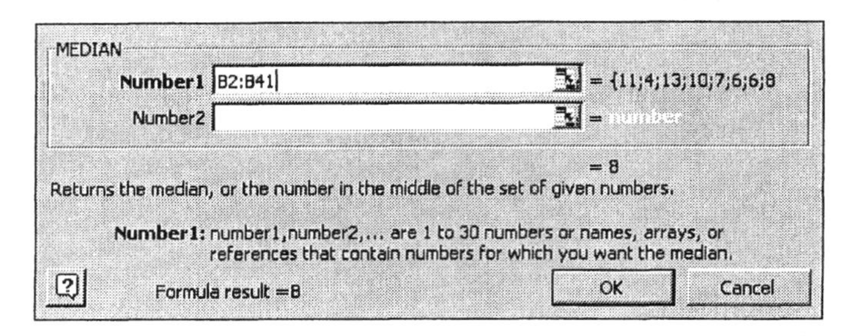

Your output should look similar to the output displayed below.

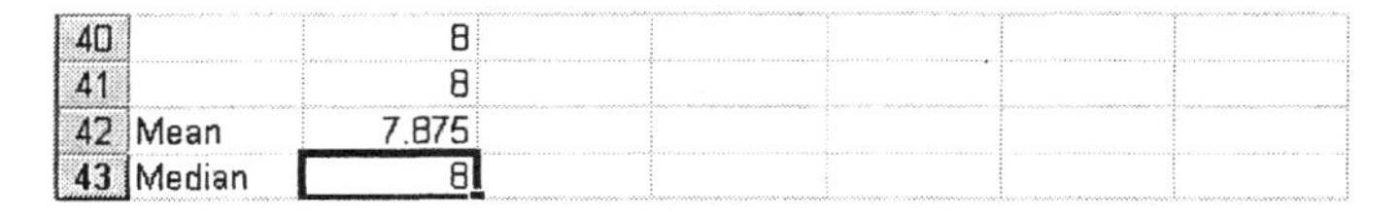

40		8
41		8
42	Mean	7.875
43	Median	8

Section 3.2 Measures of Dispersion

► Example 1 (pg. 131) Computing the Mean, Range, Variance, and Standard Deviation for a Sample

1. Open worksheet "3_2_Ex1" in the Chapter 3 folder. The first few rows are shown below.

	A	B	C	D	E	F	G
1	University	University B					
2	73	86					
3	103	91					
4	91	107					

2. At the top of the screen, click **Tools** and select **Data Analysis**.

If Data Analysis does not appear as a choice in the Tools menu, you will need to load the Microsoft Excel Analysis ToolPak add-in. Follow the procedure in Section GS 8.1 before continuing.

3. Select **Descriptive Statistics** and click **OK**.

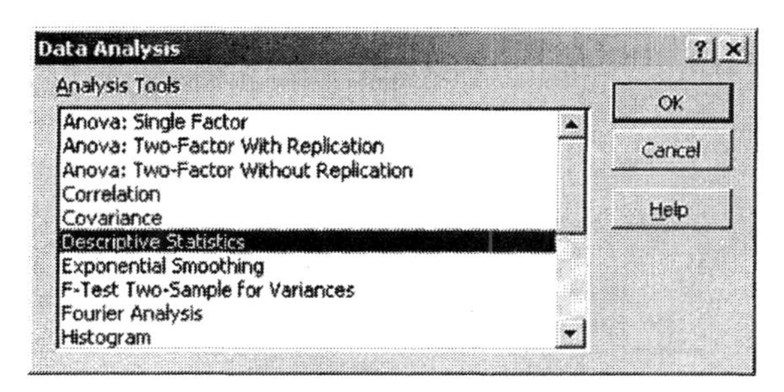

4. Complete the Descriptive Statistics dialog box as shown below. You will be obtaining descriptive statistics for the data in the worksheet range A1:B101. There are labels ("University A" and "University B") in the top cell of each column. The output will be placed in a new worksheet. Be sure to click in the **Summary statistics** box to place a checkmark there. The output will include summary statistics for both universities. Click **OK**.

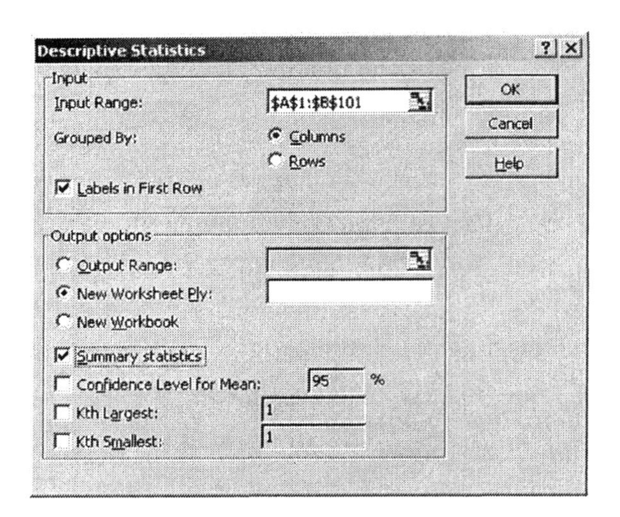

Your output will look similar to the output displayed below. I recommend that you make columns A and D wider so that you can read the complete label for each summary value. The descriptive statistics output includes the mean, standard deviation, sample variance, and the range. Note that the variance and standard deviation are computed using the unbiased estimate formula that has n – 1 in the denominator.

	A	B	C	D	E	F	G
1	University A		University B				
2							
3	Mean	100.02	Mean	99.99			
4	Standard E	1.607996	Standard E	0.834302			
5	Median	102	Median	98			
6	Mode	103	Mode	107			
7	Standard D	16.07996	Standard D	8.343019			
8	Sample Va	258.5653	Sample Va	69.60596			
9	Kurtosis	-0.03652	Kurtosis	-0.65097			
10	Skewness	0.0468	Skewness	0.340169			
11	Range	81	Range	33			
12	Minimum	60	Minimum	86			
13	Maximum	141	Maximum	119			
14	Sum	10002	Sum	9999			
15	Count	100	Count	100			

◄

▸ Example 1 (pg. 131) Constructing a Histogram

You are instructed to construct a histogram for each university's IQ score data. I will provide directions for University A only. You can then repeat the steps on your own to construct the histogram for University B.

1. Open worksheet "3_2_Ex1" in the Chapter 3 folder. The first few rows are shown below.

	A	B	C	D	E	F	G
1	University	University B					
2	73	86					
3	103	91					
4	91	107					

2. At the top of the screen, click **Tools** and select **Data Analysis**.

If Data Analysis does not appear as a choice in the Tools menu, you will need to load the Microsoft Excel Analysis ToolPak add-in. Follow the procedure in Section GS 8.1 before continuing.

3. Excel's histogram procedure uses grouped data to generate a frequency distribution and a frequency histogram. The procedure requires that you indicate a "bin" for each class. The number that you specify for each bin is actually the upper limit of the class. You are instructed to use a lower class limit of 50 for the first class and a class width of 10. The upper limit is equal to one less than the lower limit of the next higher class. For this problem, the upper limits will be 59, 69, 79, etc. By referring to the output of the previous example in this manual, you will see that the maximum value in University A is 141. So, the highest upper limit you will need 149. Enter **Bin** in cell C1 and key in the upper limits in column C as shown below.

	A	B	C	D	E	F	G
1	University	University	Bin				
2	73	86	59				
3	103	91	69				
4	91	107	79				
5	93	94	89				
6	136	105	99				
7	108	107	109				
8	92	89	119				
9	104	96	129				
10	90	102	139				
11	78	96	149				

4. Click **Tools** and select **Data Analysis**. Select **Histogram** and click **OK**.

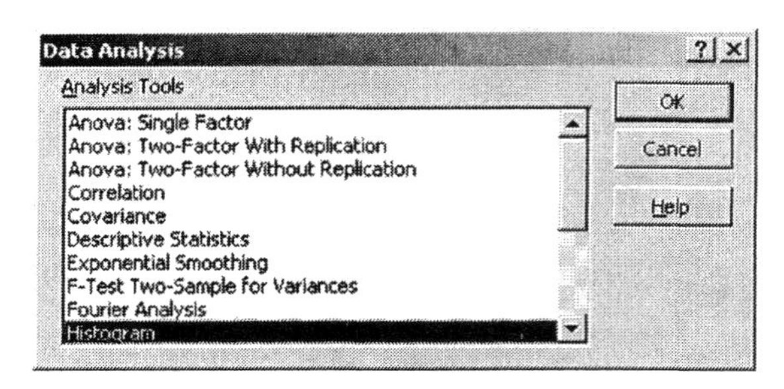

5. Complete the Histogram dialog box as shown below and click **OK**.

Note the checkmark in the box to the left of Labels. In your worksheet, "University A" appears in cell A1, and "Bin" appears in cell C1. Because you included these cells in the Input Range and Bin Range, respectively, you need to let Excel know that these cells contain labels rather than data. Otherwise Excel will attempt to use the information in these cells when constructing the frequency distribution and histogram.

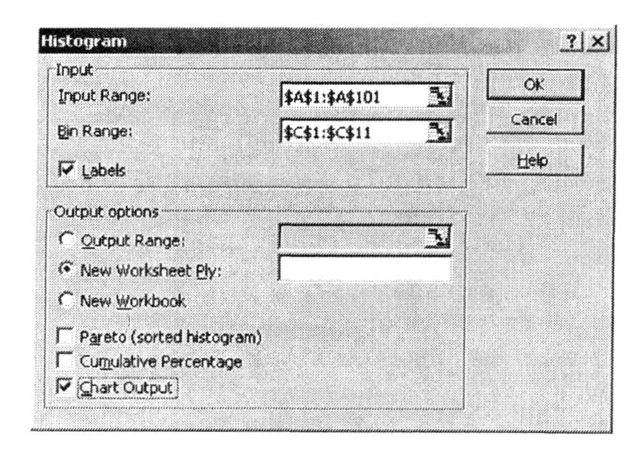

You should see output similar to the output displayed below.

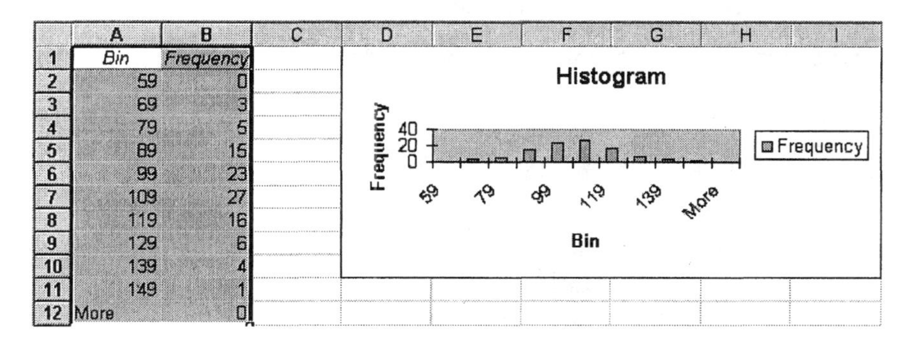

6. You will now modify this histogram so that it is displayed in a more informative manner. First, make the chart taller so that it is easier to read. To do this, click within the figure near a border. Black square handles appear. Click on the center handle on the bottom border of the figure and drag it down a few rows.

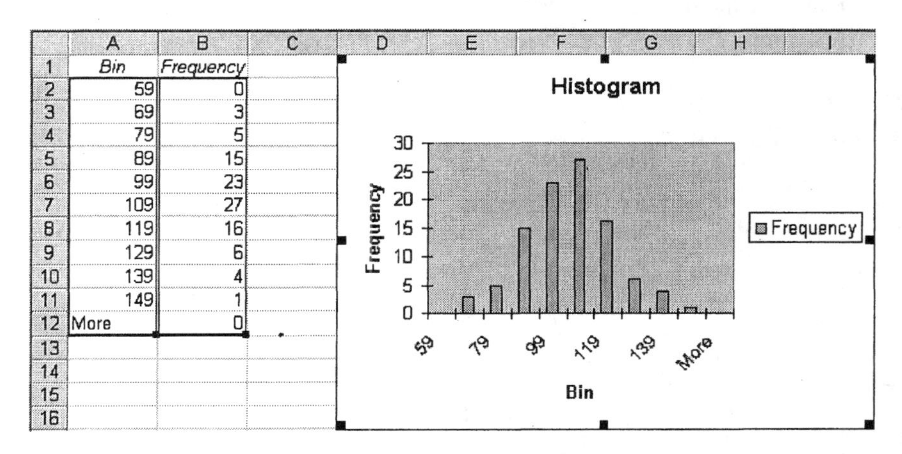

7. Next, remove the space between the vertical bars. **Right click** on one of the vertical bars. Select **Format Data Series** from the shortcut menu that appears.

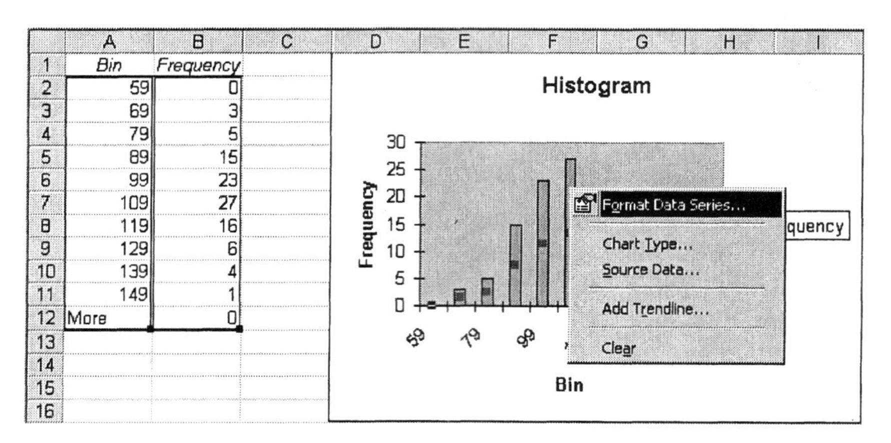

8. Click the **Options** tab at the top of the Format Data Series dialog box. Change the value in the Gap width box to **0**. Click **OK**.

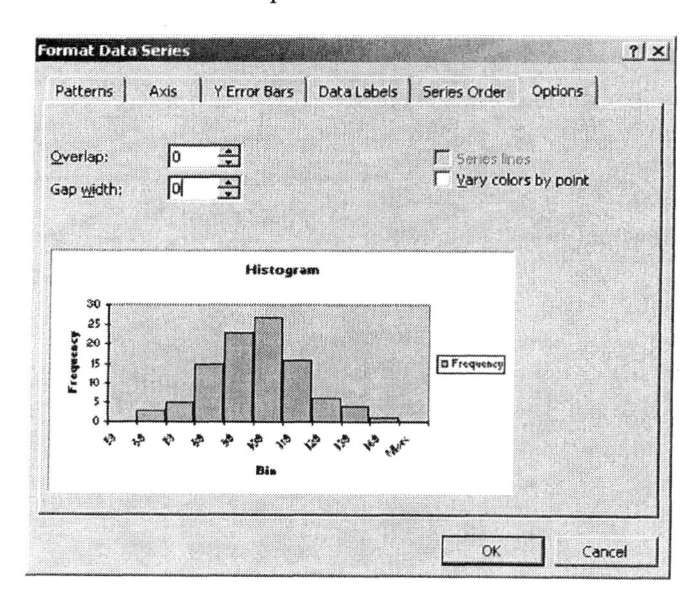

9. Change the X-axis values from upper limits to lower limits. To do this, first enter the lower limits in column C of the Excel worksheet as shown below.

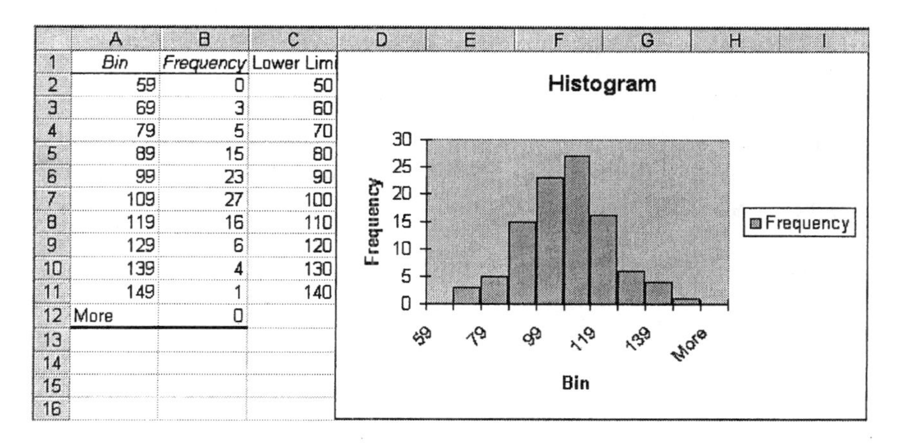

10. **Right click** on a vertical bar. Select **Source Data** from the shortcut menu that appears.

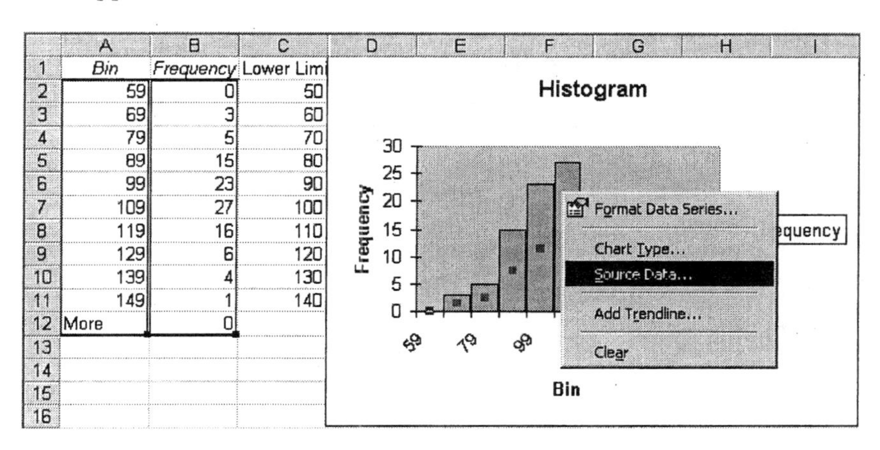

11. Click the **Series** tab at the top of the Source Data dialog box. The ranges displayed in the Values field and the Category (X) axis label field refer to the frequency distribution table in the top left of the worksheet. You do not want to include row 12, because that is the row containing information related to the "More" category. You also want the lower limit values in column C to be displayed on the X axis rather than the column A bin values. First, change the 12 to 11 in the **Values** field. Assuming that your histogram is displayed in sheet 2, the entry should read **=Sheet2!$B$2:$B$11**. Next, edit the **Category (X) axis labels** field so that the entry reads **=Sheet2!$C$2:$C$11**. Click **OK**.

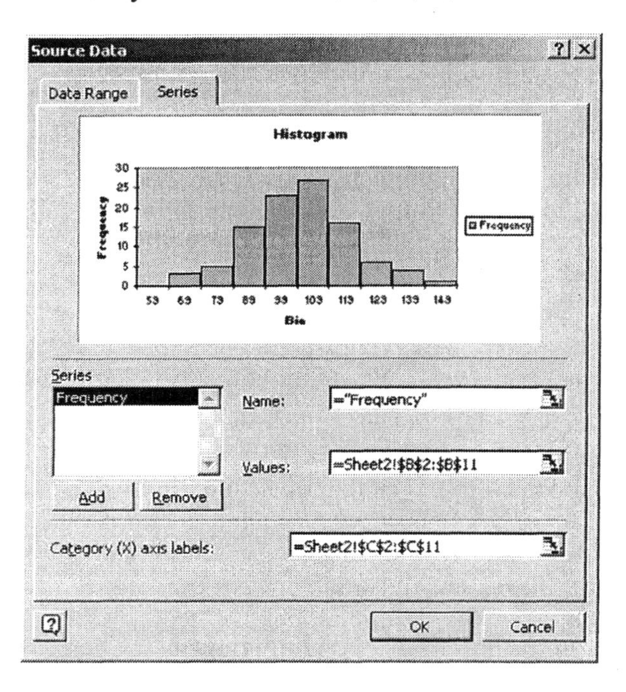

12. You will use Chart Options to modify three aspects of the histogram: Titles, gridlines, and legend. **Right click** in the gray plot area of the chart and select **Chart Options** from the shortcut menu that appears.

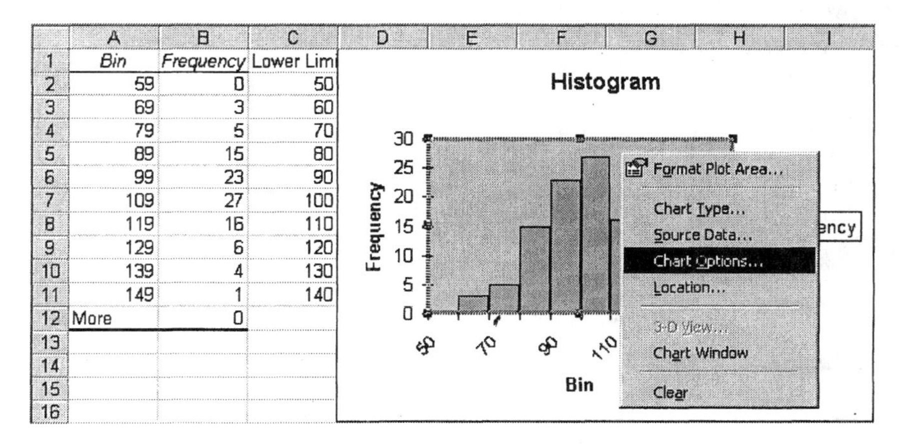

	A	B	C
1	*Bin*	*Frequency*	Lower Limi
2	59	0	50
3	69	3	60
4	79	5	70
5	89	15	80
6	99	23	90
7	109	27	100
8	119	16	110
9	129	6	120
10	139	4	130
11	149	1	140
12	More	0	

13. Click the **Titles** tab at the top of the Chart Options dialog box. Change the Chart title from "Histogram" to **University A IQ Scores**. Change the Category (X) axis label from "Bin" to **IQ Scores**.

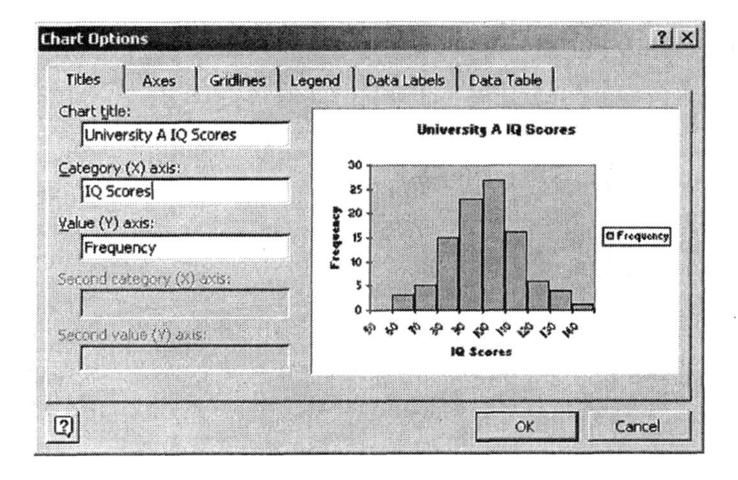

14. Click the **Gridlines** tab at the top of the Chart Options dialog box. Under Value (Y) axis, click in the **Major gridlines** box so that a checkmark appears there.

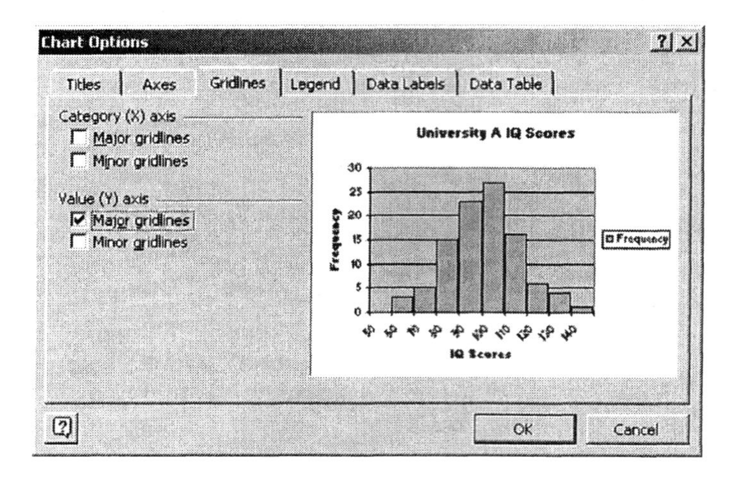

15. Click the **Legend** tab. To remove the frequency legend displayed at the right of the histogram chart, click in the **Show legend** box to remove the checkmark. Click **OK**. Your histogram chart should now appear similar to the one displayed below.

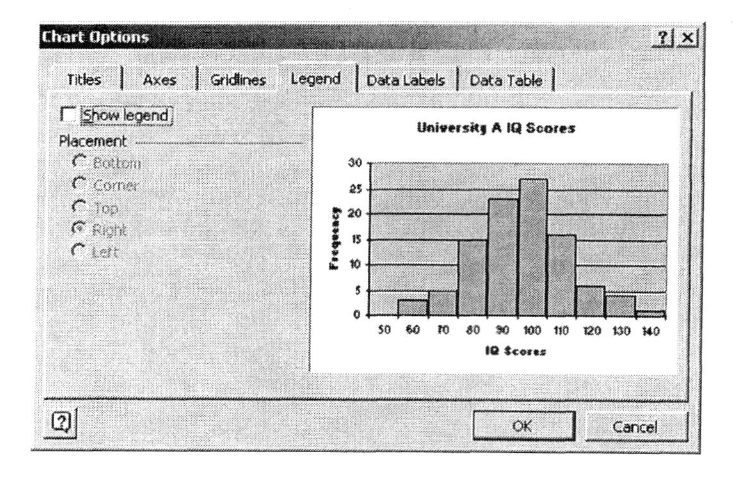

► Exercise 21 (pg. 145) Computing the Mean, Median, and Sample Standard Deviation

1. Open worksheet "3_2_21" in the Chapter 3 folder. The first few rows are shown below.

	A	B	C	D	E	F	G
1	Financial S	Energy Stocks					
2	16.12	2.49					
3	3.34	7.7					
4	28.51	1.82					

2. At the top of the screen, click **Tools** and select **Data Analysis**.

If Data Analysis does not appear as a choice in the Tools menu, you will need to load the Microsoft Excel Analysis ToolPak add-in. Follow the procedure in Section GS 8.1 before continuing.

3. Select **Descriptive Statistics** and click **OK**.

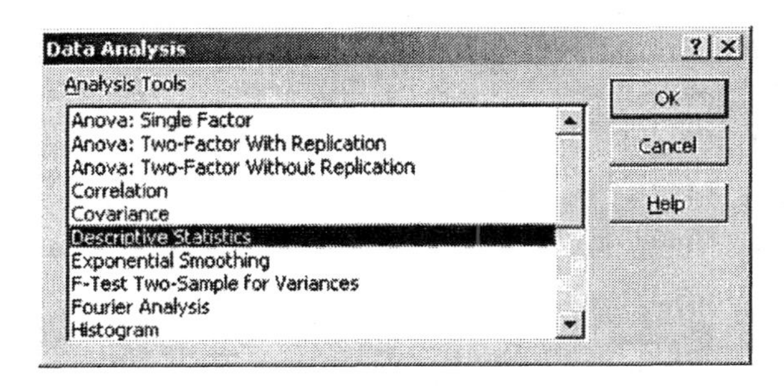

4. Complete the Descriptive Statistics dialog box as shown below. You will be obtaining descriptive statistics for the data in the worksheet range A1:B41. There are labels ("Financial Stocks" and "Energy Stocks") in the top cell of each column. The output will be placed in a new worksheet. Be sure to click in the **Summary statistics** box to place a checkmark there. The output will include summary statistics for both types of stocks. Click **OK**.

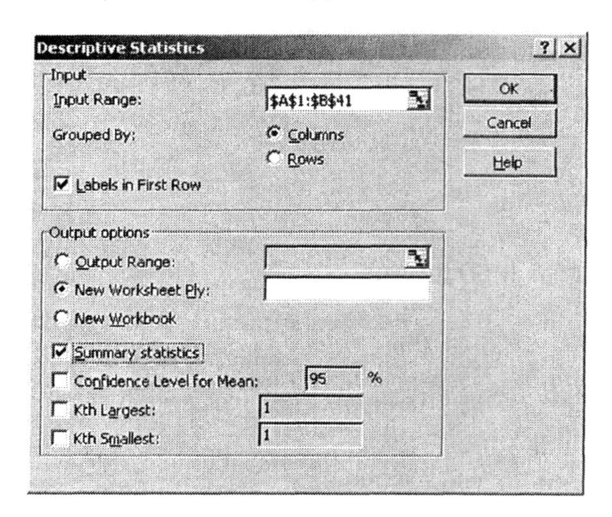

Your output will look similar to the output displayed below. I recommend that you make columns A and D wider so that you can read the complete label for each summary value

	A	B	C	D	E	F	G
1	*Financial Stocks*		*Energy Stocks*				
2							
3	Mean	11.12225	Mean	9.712571			
4	Standard Error	1.274559	Standard Error	0.989229			
5	Median	9.33	Median	9.09			
6	Mode	2.98	Mode	9.7			
7	Standard Deviation	8.061017	Standard Deviation	5.852359			
8	Sample Variance	64.98	Sample Variance	34.2501			
9	Kurtosis	0.831668	Kurtosis	4.341392			
10	Skewness	1.003895	Skewness	1.811765			
11	Range	33.59	Range	28.57			
12	Minimum	0.63	Minimum	1.82			
13	Maximum	34.22	Maximum	30.39			
14	Sum	444.89	Sum	339.94			
15	Count	40	Count	35			

Section 3.4 Measures of Position

▶ Example 2 (pg. 160) Determining the Data Value of a Percentile

1. Open worksheet "3_4_Ex2" in the Chapter 3 folder. The first few rows of the worksheet are displayed below.

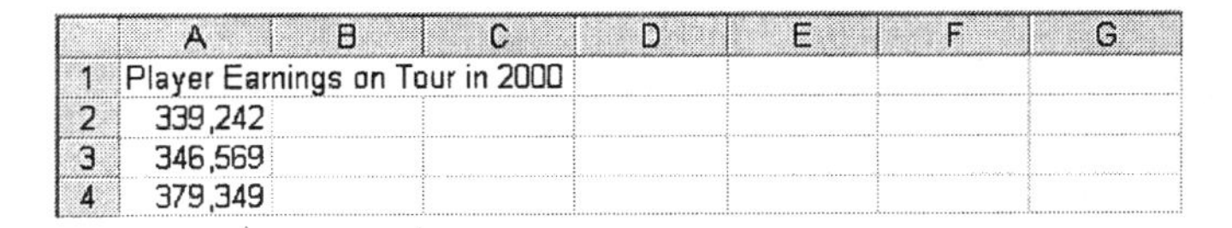

	A	B	C	D	E	F	G
1	Player Earnings on Tour in 2000						
2	339,242						
3	346,569						
4	379,349						

2. Click in the cell where you would like to display the output. I clicked in cell **D1**.

3. At the top of the screen, click **Insert** and select **Function**.

4. Under Function category, select **Statistical**. Under Function name, select **PERCENTILE**. Click **OK**.

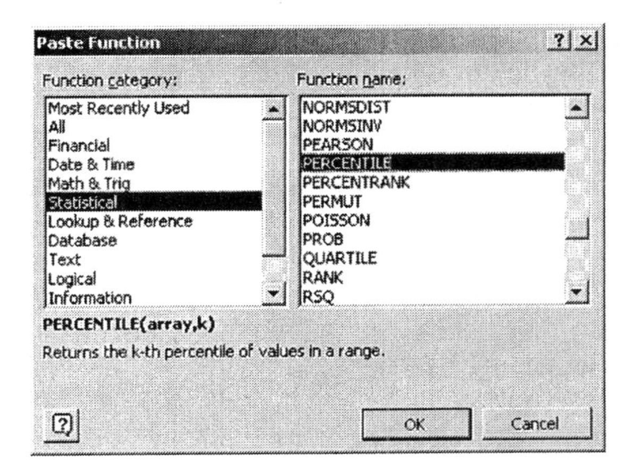

5. Complete the Percentile dialog box as shown below. Array refers to the worksheet range of the player earnings data. K refers to the percentile expressed as a proportion (e.g., .85). Click **OK**.

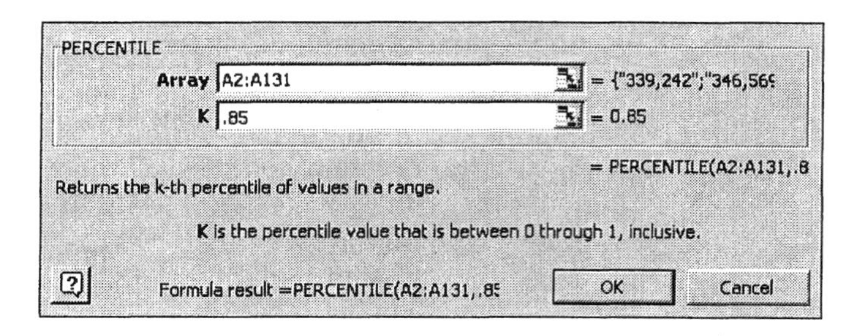

The output is displayed below. The 85th percentile is $1,814,112.

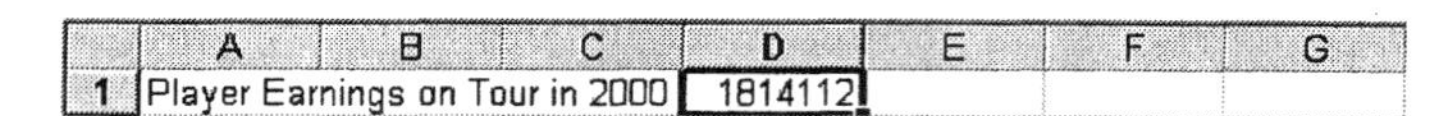

	A	B	C	D	E	F	G
1	Player Earnings on Tour in 2000			1814112			

► Exercise 11 (pg. 166) Computing Z-scores, Quartiles, Interquartile Range, and Fences

You will use the red blood cell count data to learn how to compute z-scores, quartiles, the interquartile range, and fences.

1. Open worksheet "3_4_11" in the Chapter 3 folder. The first few rows are displayed below.

	A	B	C	D	E	F	G
1	Red Blood Cell Count						
2	6.2						
3	6.8						
4	6.1						

2. Begin by computing the mean and standard deviation for the distribution. You will need these values to compute z-scores. First, insert a column to the left of the data where you will place labels for your output. At the top of the screen, click **Insert** and select **Columns**.

3. Enter the labels **Mean** and **StDev** in column A as shown below.

21		7.5					
22	Mean						
23	StDev						

4. Click in cell **B22** where you will place the mean.

5. At the top of the screen, click **Insert** and select **Function**.

6. Under Function category, select **Statistical**. Under Function name, select **AVERAGE**. Click **OK**.

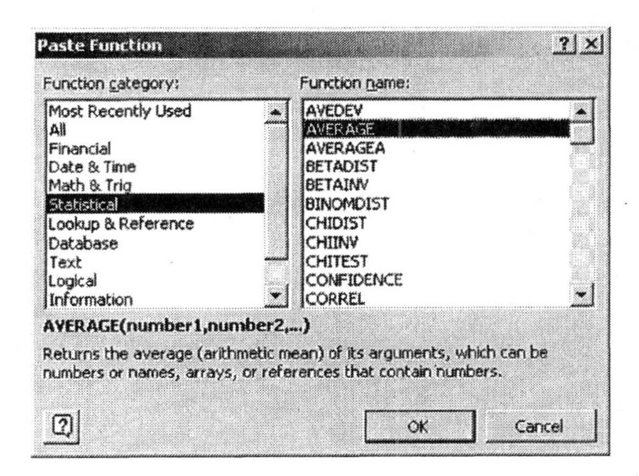

7. You should see the range B2:B21 in the Number 1 window of the dialog box. If this range does not appear, you will need to enter it. Click **OK**.

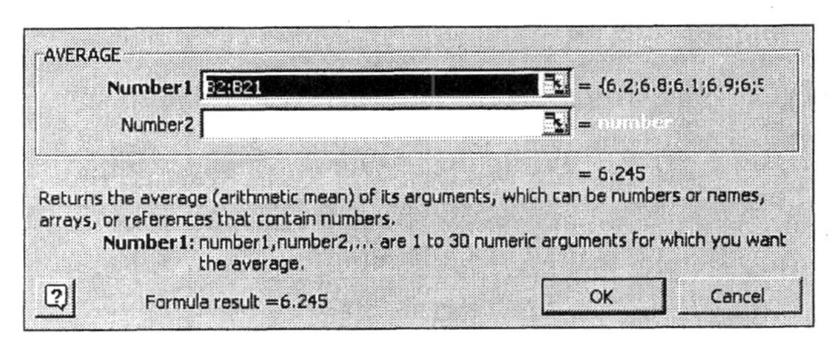

8. You will see 6.245 in cell B22. Next, click in **B23** where you will place the sample standard deviation.

9. At the top of the screen, click **Insert** and select **Function**.

10. Under Function category, select **Statistical**. Under Function name, select **STDEV**. Click **OK**.

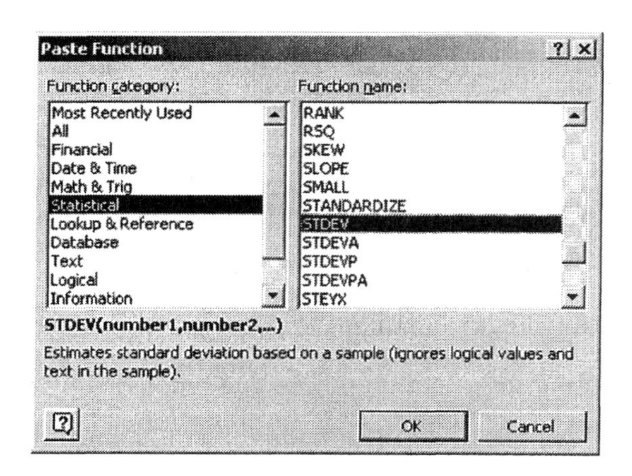

11. The range B2:B22 will automatically appear in the Number 1 window. You will need to change the range to B2:B21 so that the mean is not included in the calculations. Click **OK**.

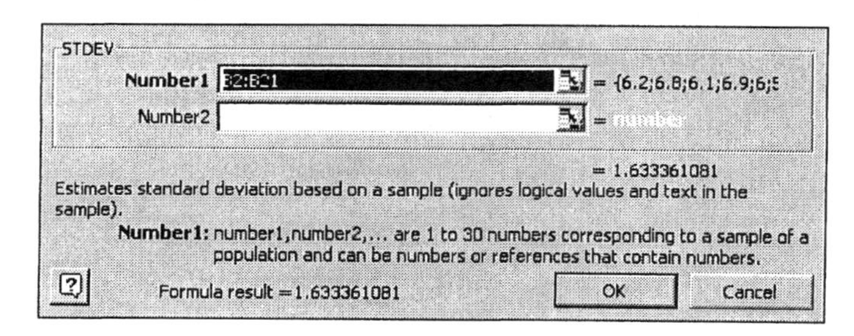

12. The STDEV function returns a value of 1.6334.

13. Enter the label **z-score** in cell D1 of the worksheet. Then click in cell **D2** where you will place the z-score for 6.2.

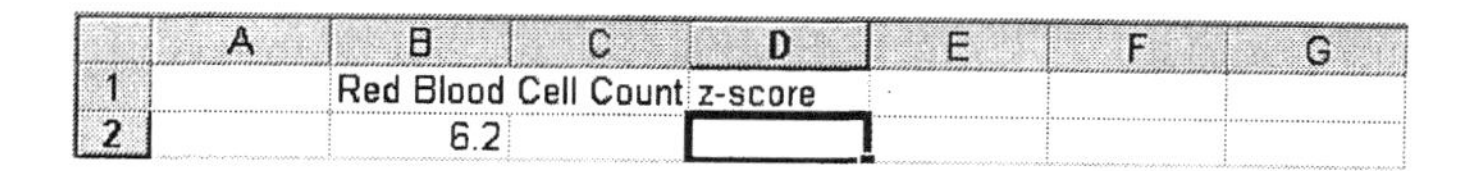

14. At the top of the screen click **Insert** and select **Function**.

15. Under Function category select **Statistical**. Under Function name, select **STANDARDIZE**. Click **OK**.

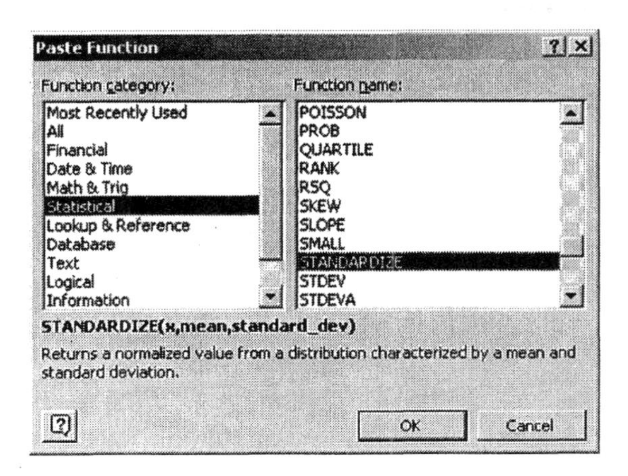

16. Complete the Standardize dialog box as shown below. You are requesting the z-score for the red blood cell count value in cell B2 of the worksheet. The mean is located in cell $B$22 of the worksheet. Note that the dollar signs are necessary here because you want the cell location of the mean to be an absolute reference. This is because you will be copying the function to other cells in the worksheet. The sample standard deviation is located in cell $B$23 of the worksheet. The cell reference of the standard deviation must also be an absolute reference. Click **OK**.

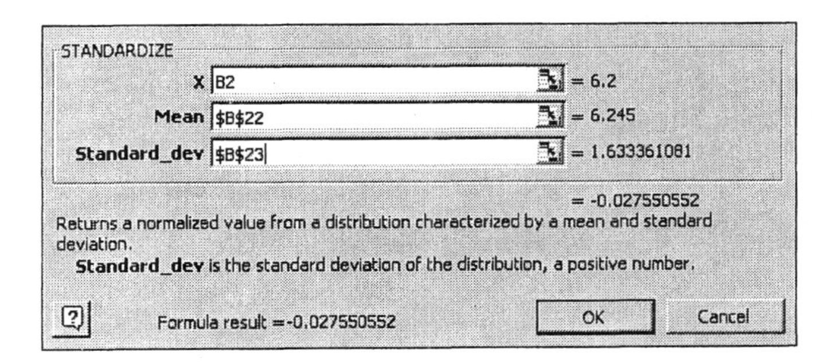

17. To obtain z-scores for all values in the distribution, copy the contents of cell D2 to cells D3 through D21. The first ten rows of the worksheet are displayed below. The z-score for a count of 0.2 is –3.7010.

	A	B	C	D	E	F	G
1		Red Blood Cell Count		z-score			
2		6.2		-0.02755			
3		6.8		0.33979			
4		6.1		-0.08877			
5		6.9		0.401014			
6		6		-0.15			
7		5.2		-0.63979			
8		6.1		-0.08877			
9		0.2		-3.70096			
10		8.1		1.135695			

18. Next, you will find the quartiles and the interquartile range (IQR) for this distribution. Key in labels for the quartiles as shown below. Then click in cell **G1** to place the first quartile there.

	A	B	C	D	E	F	G
1		Red Blood Cell Count		z-score		Quartile 1	
2		6.2		-0.02755		Quartile 2	
3		6.8		0.33979		Quartile 3	
4		6.1		-0.08877		IQR	

19. You will be using the QUARTILE function to obtain the first, second, and third quartiles. At the top of the screen, click **Insert** and select **Function.**

20. Under Function category, select **Statistical**. Under Function name, select **QUARTILE**. Click **OK**.

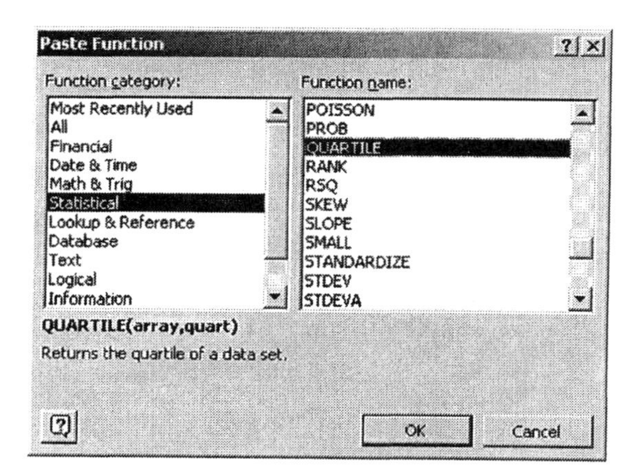

21. Complete the Quartile dialog box as shown below. Click **OK**.

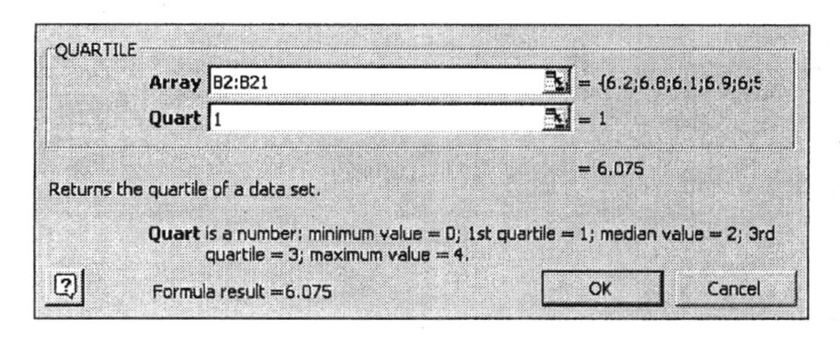

22. Click in cell **G2** to place the second quartile there.

23. At the top of the screen, click **Insert** and select **Function**.

24. Under Function category, select **Statistical**. Under Function name, select **QUARTILE**. Click **OK**.

25. Complete the Quartile dialog box as shown below. Click **OK**.

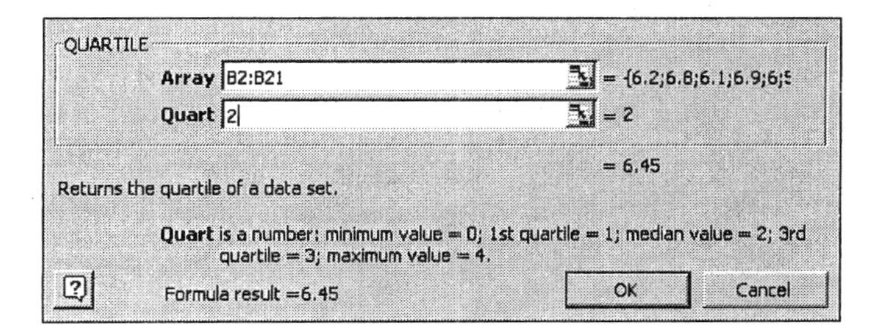

26. Click in cell **G3** to place the third quartile there.

27. At the top of the screen, click **Insert** and select **Function**.

28. Under Function category, select **Statistical**. Under Function name, select **QUARTILE**. Click **OK**.

29. Complete the Quartile dialog box as shown below. Click **OK**.

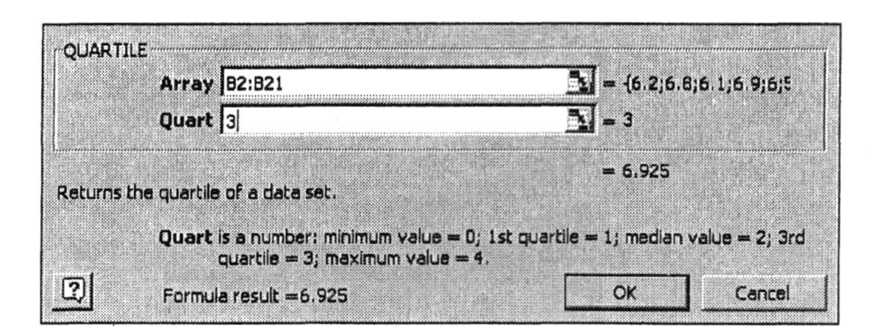

30. Click in cell **G4** where you will place the IQR.

31. You will obtain the IQR by subtracting Quartile 1 from Quartile 3. Enter **=G3-G1** in cell G4. Then press [**Enter**].

	A	B	C	D	E	F	G
1		Red Blood Cell Count		z-score		Quartile 1	6.075
2		6.2		-0.02755		Quartile 2	6.45
3		6.8		0.33979		Quartile 3	6.925
4		6.1		-0.08877		IQR	=G3-G1

The IQR for this distribution is .85.

	A	B	C	D	E	F	G
1		Red Blood Cell Count		z-score		Quartile 1	6.075
2		6.2		-0.02755		Quartile 2	6.45
3		6.8		0.33979		Quartile 3	6.925
4		6.1		-0.08877		IQR	0.85

32. The formulas for the Lower Fence and Upper Fence are provided on page 163 of your textbook. First enter the labels as shown below. Then click in cell **G6** to place the formula for the Upper Fence there.

	A	B	C	D	E	F	G
1		Red Blood Cell Count		z-score		Quartile 1	6.075
2		6.2		-0.02755		Quartile 2	6.45
3		6.8		0.33979		Quartile 3	6.925
4		6.1		-0.08877		IQR	0.85
5		6.9		0.401014			
6		6		-0.15		Lower Fence	
7		5.2		-0.63979		Upper Fence	

33. Enter **=G1-1.5*G4** in cell G6. Press **[Enter]**.

	A	B	C	D	E	F	G	H
1		Red Blood Cell Count		z-score		Quartile 1	6.075	
2		6.2		-0.02755		Quartile 2	6.45	
3		6.8		0.33979		Quartile 3	6.925	
4		6.1		-0.08877		IQR	0.85	
5		6.9		0.401014				
6		6		-0.15		Lower Fen	=G1-1.5*G4	
7		5.2		-0.63979		Upper Fence		

34. Enter = **G3+1.5*G4** in cell G7. Press **[Enter]**.

	A	B	C	D	E	F	G	H
1		Red Blood Cell Count		z-score		Quartile 1	6.075	
2		6.2		-0.02755		Quartile 2	6.45	
3		6.8		0.33979		Quartile 3	6.925	
4		6.1		-0.08877		IQR	0.85	
5		6.9		0.401014				
6		6		-0.15		Lower Fen	4.8	
7		5.2		-0.63979		Upper Fen	=G3+1.5*G4	

Your output should look similar to the output displayed below.

	A	B	C	D	E	F	G
1		Red Blood Cell Count		z-score		Quartile 1	6.075
2		6.2		-0.02755		Quartile 2	6.45
3		6.8		0.33979		Quartile 3	6.925
4		6.1		-0.08877		IQR	0.85
5		6.9		0.401014			
6		6		-0.15		Lower Fen	4.8
7		5.2		-0.63979		Upper Fen	8.2

◄

Section 3.5 The Five-Number Summary; Boxplots

► Exercise 3 (pg. 173) **Finding the Five-Number Summary and Constructing a Boxplot**

You will use the inauguration data to learn how to find the five-number summary and construct a boxplot.

If the PHStat add-in has not been loaded, you will need to load it before continuing. Follow the instructions in Section GS 8.2.

---

1. Open worksheet "3_5_3" in the Chapter 3 folder. The first few lines of the worksheet are shown below.

	A	B	C	D	E	F	G
1	Age of Presidents						
2	57						
3	54						
4	65						

2. At the top of the screen, click **PHStat** and select **Descriptive Statistics → Box-and-Whisker Plot**.

3. Complete the Box-and-Whisker Plot dialog box as shown below. To enter the cell range, first click in the Raw Data Cell Range Window of the dialog box. Then click and drag over the range A1 through A44 of the worksheet to enter the range. Or you could simply key in A1:A44. Click **OK**.

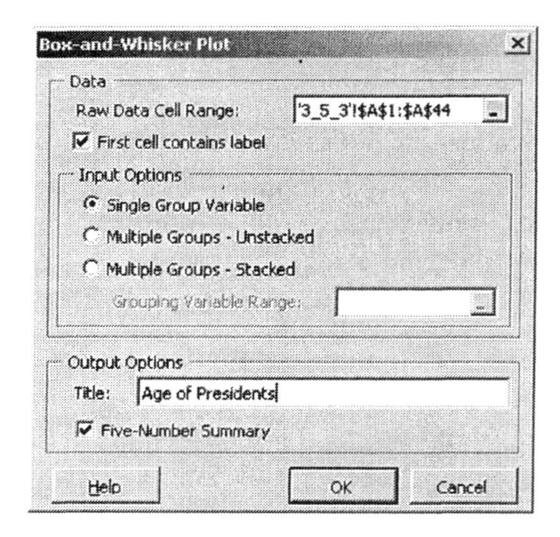

The Five-Number Summary is displayed in a worksheet named "Five Numbers."

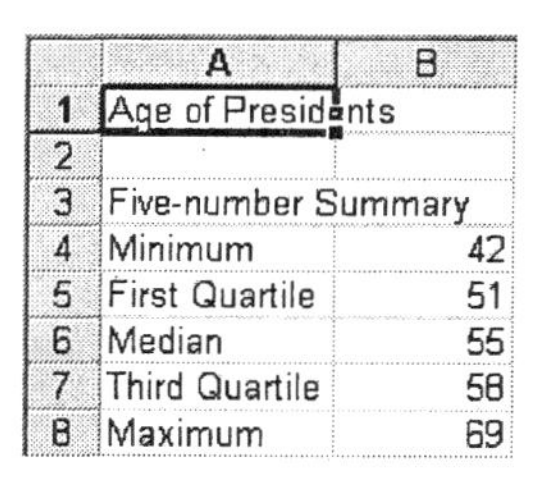

	A	B
1	Age of Presidents	
2		
3	Five-number Summary	
4	Minimum	42
5	First Quartile	51
6	Median	55
7	Third Quartile	58
8	Maximum	69

The box-and-whisker plot is displayed in a worksheet named "BoxWhiskerPlot."

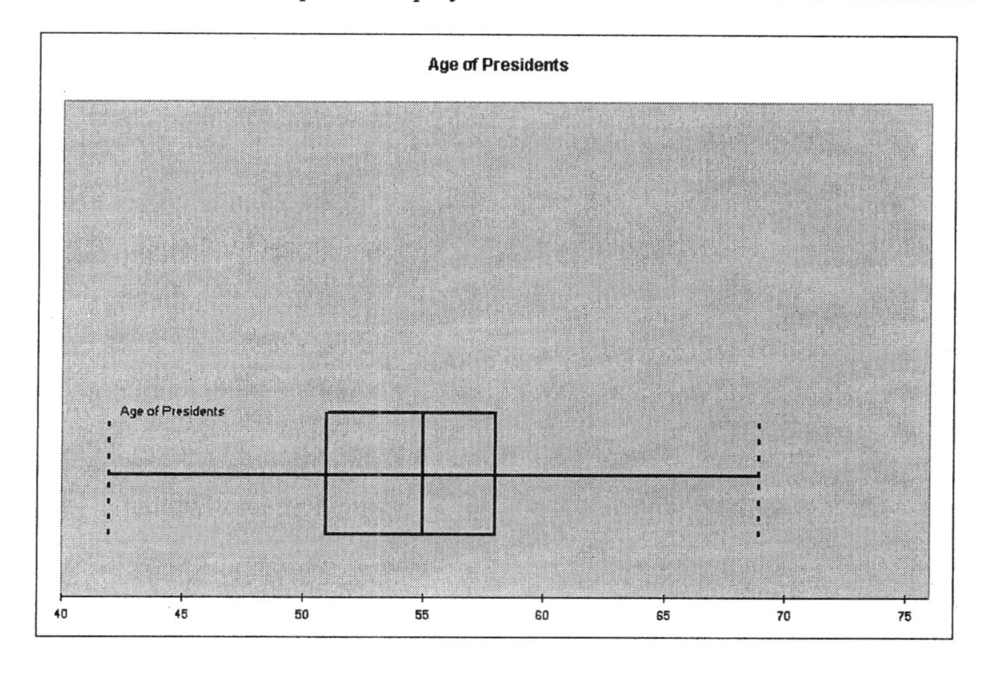

◀

Describing the Relation between Two Variables

CHAPTER 4

Section 4.1 Scatter Diagrams; Correlation

► Example 1 (pg. 193) **Drawing a Scatter Diagram**

1. Open worksheet "4_1_Ex1" in the Chapter 4 folder. The first few rows are shown below.

	A	B	C	D	E	F	G
1	Country	Per Capita	Life Expectancy				
2	Austria	21.4	77.48				
3	Belgium	23.2	77.53				
4	Finland	20	77.32				

2. Click in any cell of the data table. Then, at the top of the screen, click **Insert** and select **Chart**.

3. Under Chart type, select **XY (Scatter)**. Under Chart sub-type, select the topmost diagram. Click **Next>**.

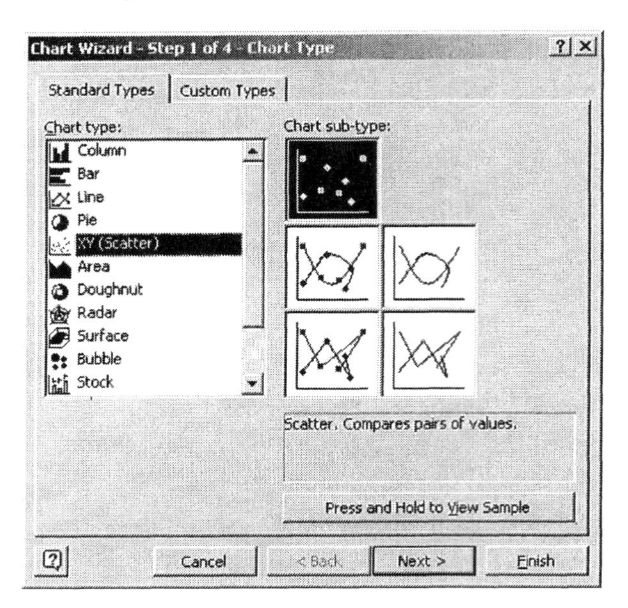

4. Change the data range as shown below so that only columns B and C are included. Click **Next>**.

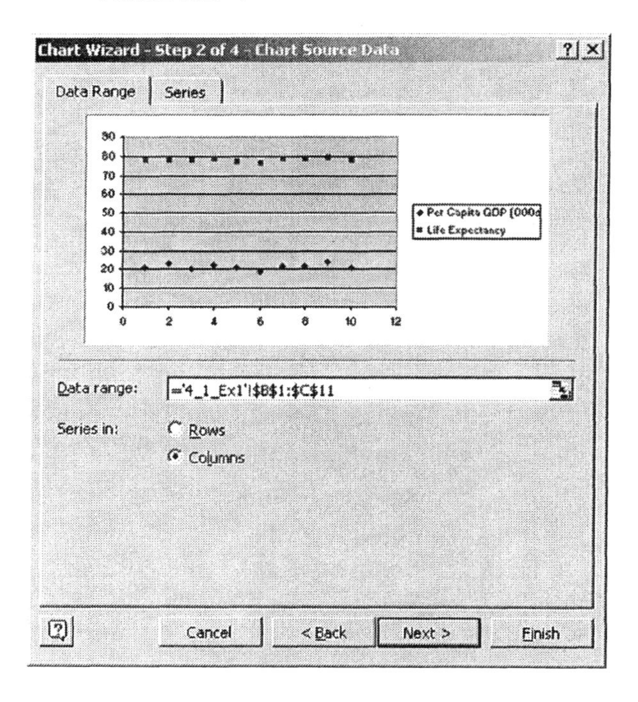

5. Click the **Titles** tab at the top of the Chart Options dialog box. Enter a title for the chart and labels for the X and Y axes. Chart title: **Scatter Diagram of Per Capita GDP and Life Expectancy**. Value (X) axis: **GDP (000s)**. Value (Y) axis: **Life Expectancy**.

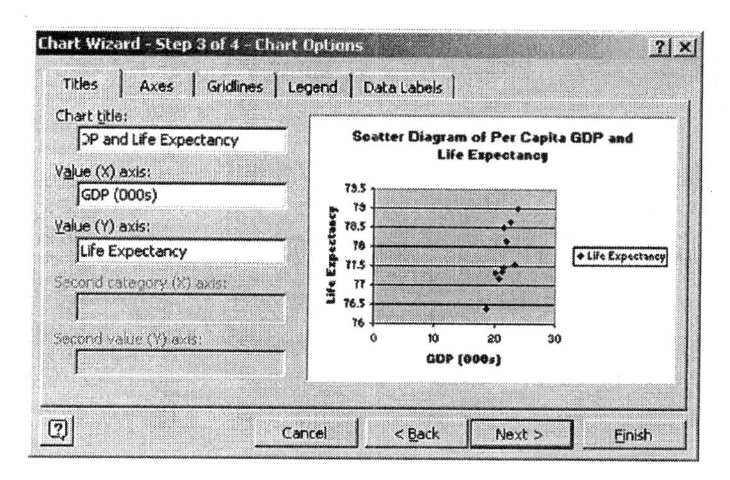

6. Click the **Gridlines** tab at the top of the Chart Options dialog box. Under Value (X) axis, click in the box to the left of **Major gridlines** to place a checkmark there.

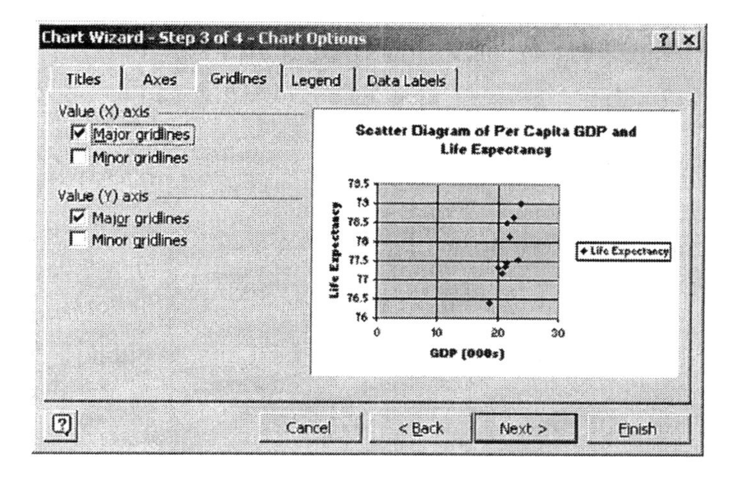

7. Click the **Legend** tab at the top of the Chart Options dialog box. Click in the box to the left of **Show legend** to remove the checkmark. Click **Next>**.

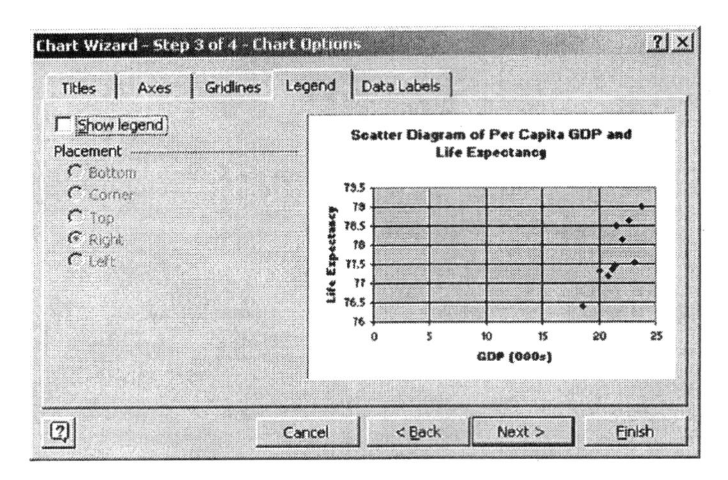

8. In the Chart Location dialog box, select **As new sheet**. Click **Finish**.

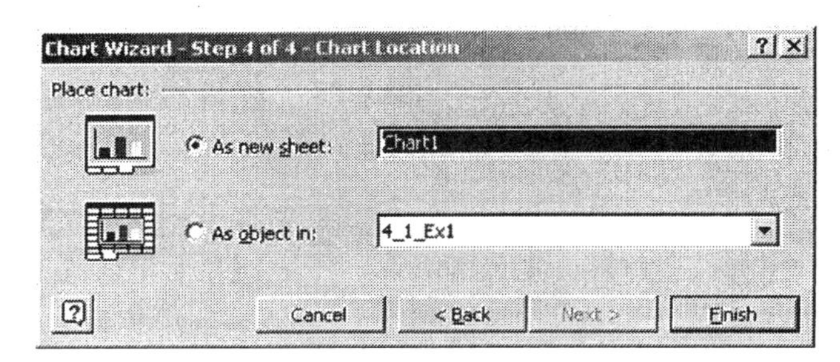

9. There is a large amount of unused space to the left of the dots. Change the X-axis so that it begins at 15 rather than 0. To do this, **right click** directly on one of the vertical lines in the chart. Select **Format Gridlines** from the shortcut menu that appears.

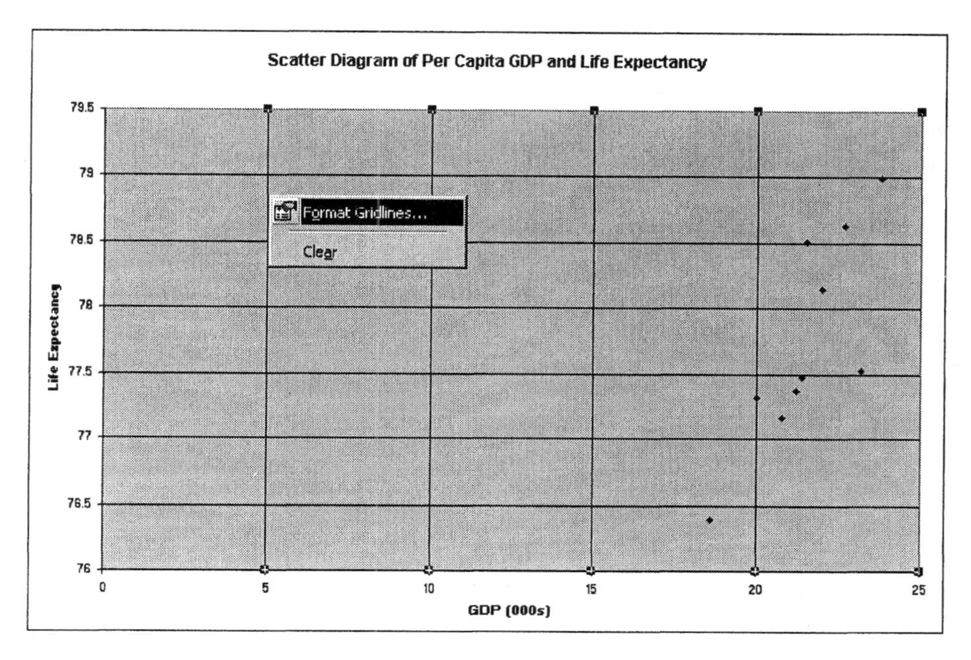

10. Click the **Scale** tab at the top of the dialog box. Change the Minimum from 0 to **15**. Click **OK**.

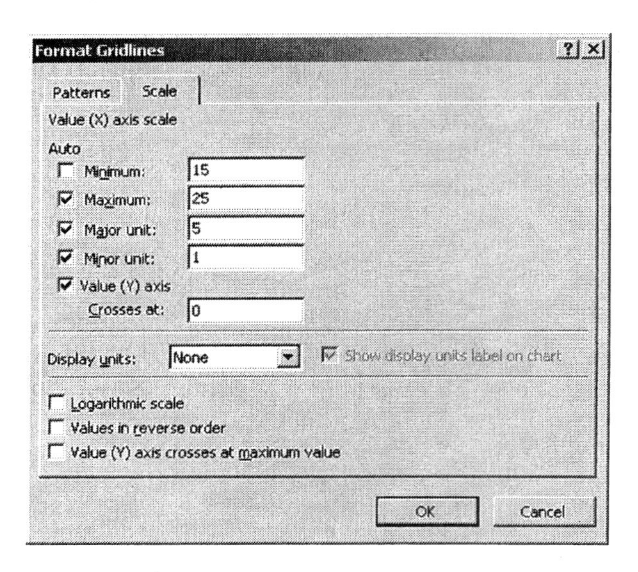

11. The dots in the scatter diagram are very small and hard to see. If you would like to make them larger, first **right click** directly on one of the dots. Then select **Format Data Series** from the shortcut menu that appears.

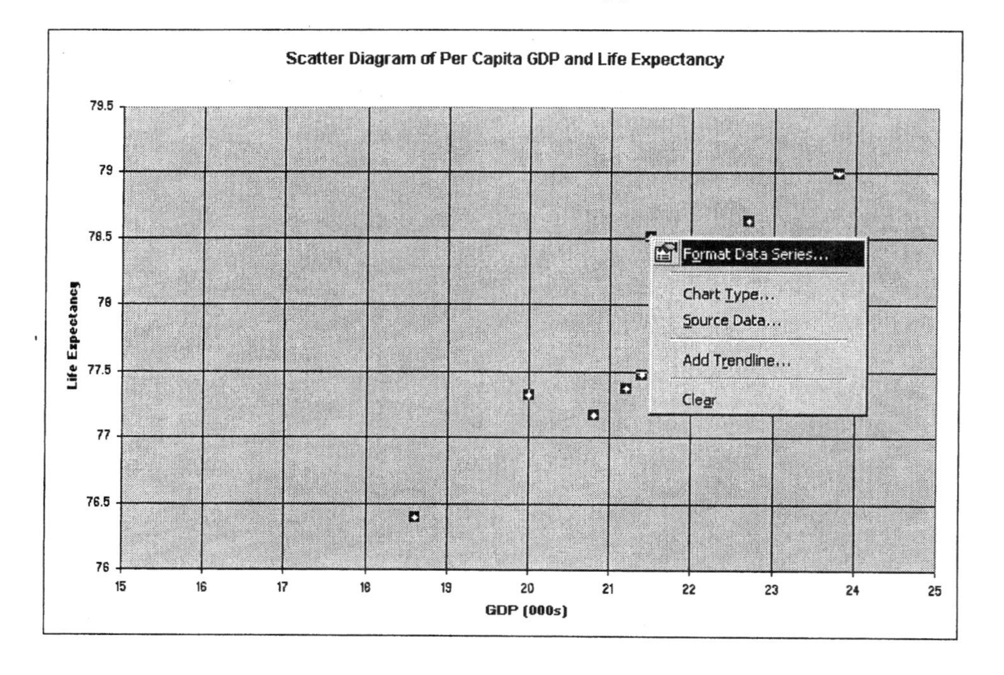

12. Click the **Patterns** tab at the top of the dialog box. You can make a size adjustment in the lower right of the Format Data Series dialog box. Change the size from 5 to **10** pts. Click **OK**.

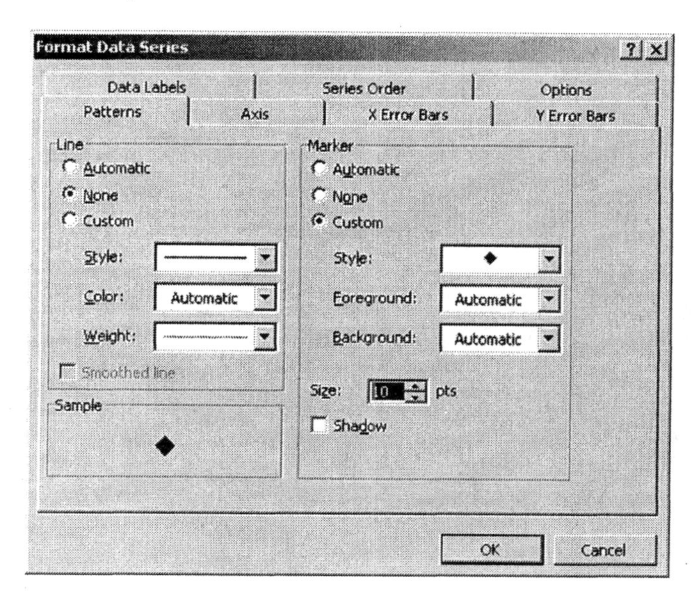

The completed scatter diagram should look similar to the one displayed below.

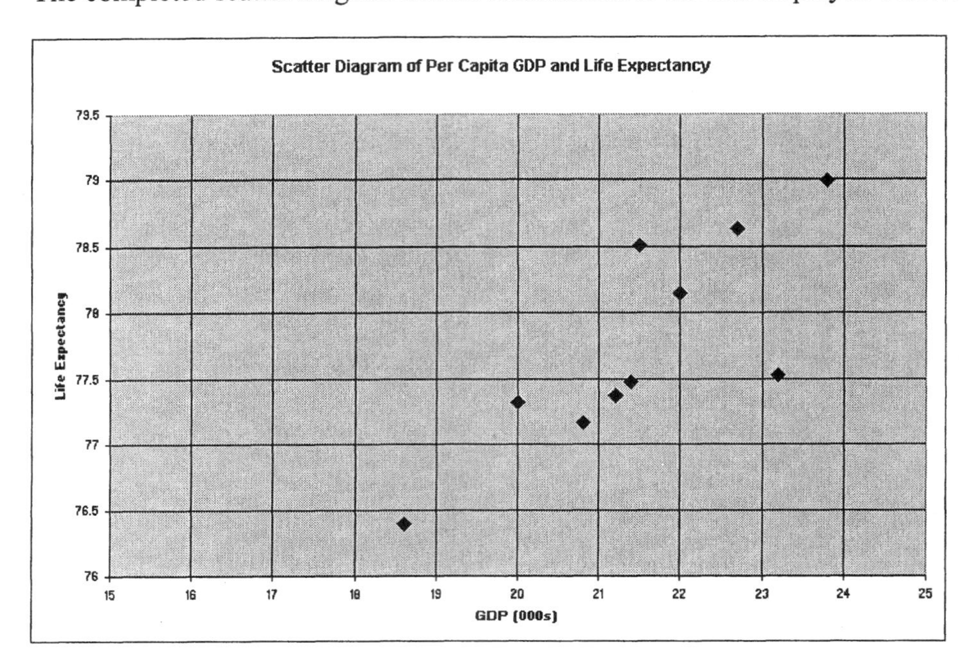

◀

▶ Example 1 (pg. 193) Computing the Correlation Coefficient

1. Open worksheet "4_1_Ex1" in the Chapter 4 folder. The first few rows are shown below.

	A	B	C	D	E	F	G
1	Country	Per Capita	Life Expectancy				
2	Austria	21.4	77.48				
3	Belgium	23.2	77.53				
4	Finland	20	77.32				

2. At the top of the screen, click **Tools** and select **Data Analysis**.

If Data Analysis does not appear as a choice in the Tools menu, you will need to load the Microsoft Excel Analysis ToolPak add-in. Follow the procedure in Section GS 8.1 before continuing.

---

3. In the Data Analysis dialog box, select **Correlation** and click **OK**.

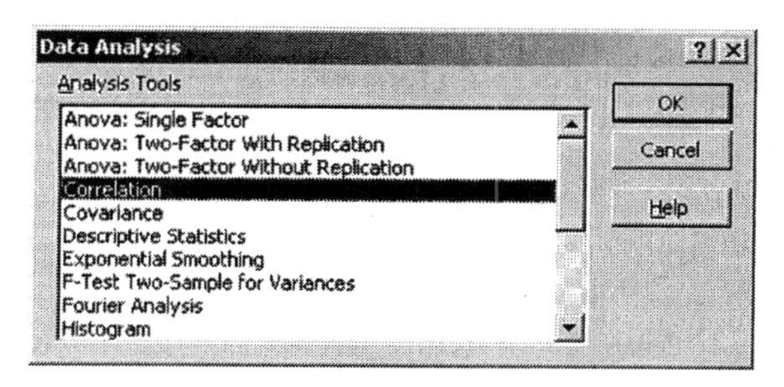

4. Complete the fields in the Correlation dialog box as shown below. Click **OK**.

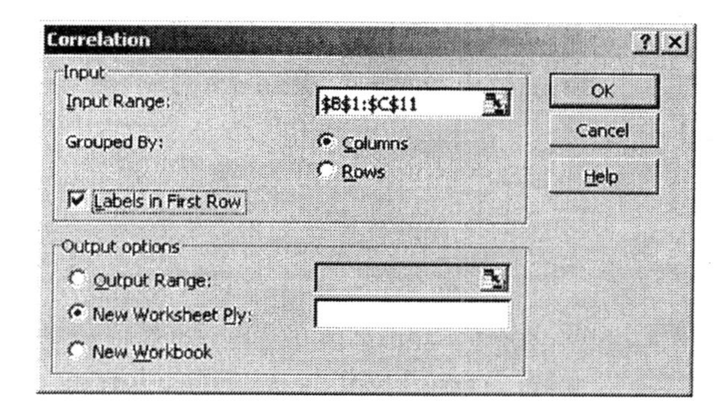

The output will be a correlation matrix that looks similar to the one shown below. The correlation between GDP and life expectancy is 0.8094.

	A	B	C	D	E	F	G
1		Per Capita GDP	e Expectancy				
2	Per Capita	1					
3	Life Expec	0.80942	1				
4							

◀

Section 4.2 Least-Squares Regression

► Exercise 9 (pg. 216)

Finding the Least-Squares Regression Line

1. Open worksheet "4_2_9" in the Chapter 4 folder. The first few rows are shown below.

	A	B	C	D	E	F	G
1	Height (inc	Head Circumference (inches)					
2	27.75	17.5					
3	24.5	17.1					
4	25.5	17.1					

2. At the top of the screen, click **Tools** and select **Data Analysis**.

If Data Analysis does not appear as a choice in the Tools menu, you will need to load the Microsoft Excel Analysis ToolPak add-in. Follow the procedure in Section GS 8.1 before continuing.

3. In the Data Analysis dialog box, select **Regression** and click **OK**.

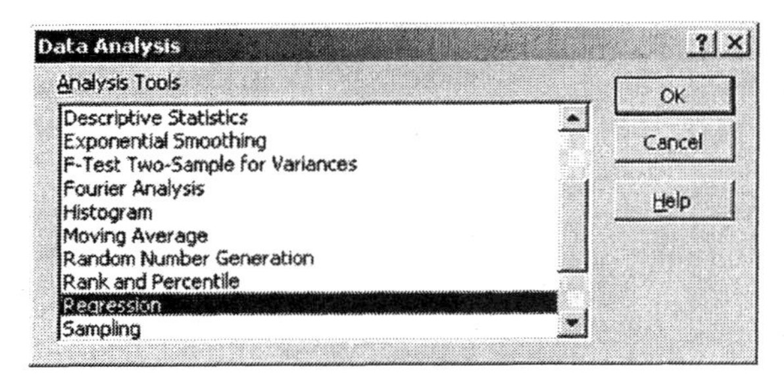

4. Complete the Regression dialog box as shown below. Click **OK**.

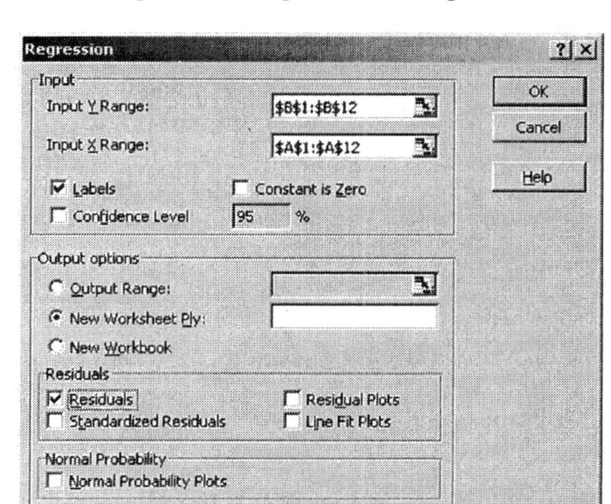

Because the output is lengthy, I will display it in two parts. The Multiple R, 0.9111, near the top of the output is the same as r for bivariate regression analysis. The intercept, shown near the bottom, is 12.4932. The slope, immediately below the intercept, is 0.1827.

	A	B	C	D	E	F	G
1	SUMMARY OUTPUT						
2							
3	*Regression Statistics*						
4	Multiple R	0.911073					
5	R Square	0.830054					
6	Adjusted R	0.811171					
7	Standard E	0.095384					
8	Observatio	11					
9							
10	ANOVA						
11		*df*	*SS*	*MS*	*F*	*ignificance F*	
12	Regressior	1	0.399935	0.399935	43.95785	9.59E-05	
13	Residual	9	0.081883	0.009098			
14	Total	10	0.481818				
15							
16		*Coefficients*	*tandard Err*	*t Stat*	*P-value*	*Lower 95%*	*Upper 95%*
17	Intercept	12.49317	0.729685	17.12132	3.56E-08	10.84251	14.14383
18	Height (inc	0.182732	0.027561	6.630072	9.59E-05	0.120385	0.24508

The output section entitled RESIDUAL OUTPUT presents predicted circumference for each height value in the data set. For example, the first height value is 27.75. The predicted circumference for that height is 17.5640. The Residuals column presents the difference between the observed height and predicted height. The observed height for the first observation is 17.5. The difference between 17.5 and 17.5640 is –0.0640.

22	RESIDUAL OUTPUT		
23			
24	Observation	Predicted Circumfe	Residuals
25	1	17.56399	-0.06399
26	2	16.97011	0.129886
27	3	17.15285	-0.05285
28	4	17.24421	0.055787
29	5	17.06148	-0.16148
30	6	17.56399	0.036006
31	7	17.33558	-0.03558
32	8	17.42694	0.073055
33	9	17.38126	-0.08126
34	10	17.38126	0.118738
35	11	17.51831	-0.01831

◀

Section 4.3 Diagnostics on the Least-Squares Regression Line

► Exercise 11 (pg. 232)

Drawing a Scatter Diagram and a Residual Plot

You will draw a scatter diagram and a residual plot using the Other Old Faithful data.

1. Open worksheet "4_3_11" in the Chapter 4 folder. The first few rows are shown below.

	A	B	C	D	E	F	G
1	Time Betw	Length					
2	12.17	1.88					
3	11.63	1.77					
4	12.03	1.83					

2. You will begin by drawing the scatter diagram. Click in any cell of the data table. Then, at the top of the screen, click **Insert** and select **Chart**.

3. Under Chart type, select **XY (Scatter)**. Under Chart sub-type, select the topmost diagram. Click **Next>**.

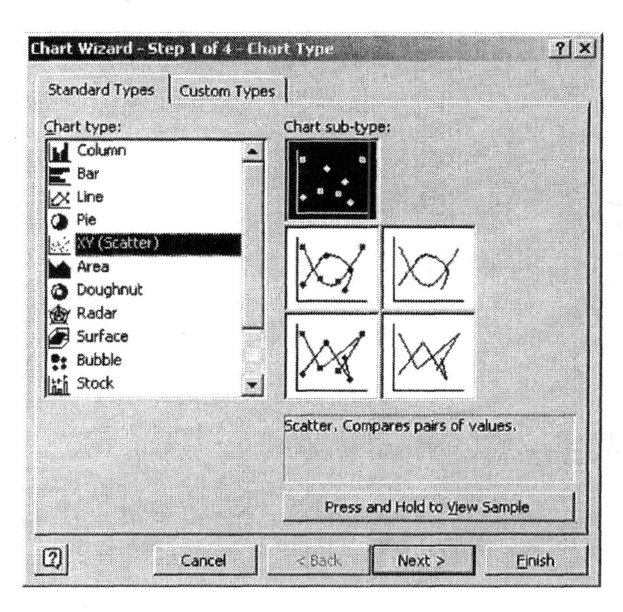

4. The data range should be the same as shown below. Make any necessary corrections. Then click **Next>**.

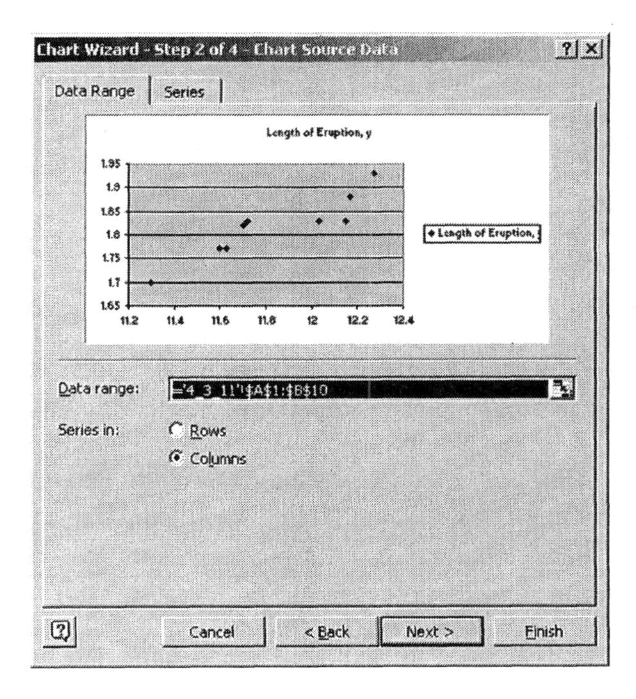

5. Enter a title for the chart and labels for the X and Y axes. Chart title: **Scatter Diagram of Time between Eruptions and Length of Eruptions**. Value (X) axis: **Time between Eruptions**. Value (Y) axis: **Length of Eruptions**.

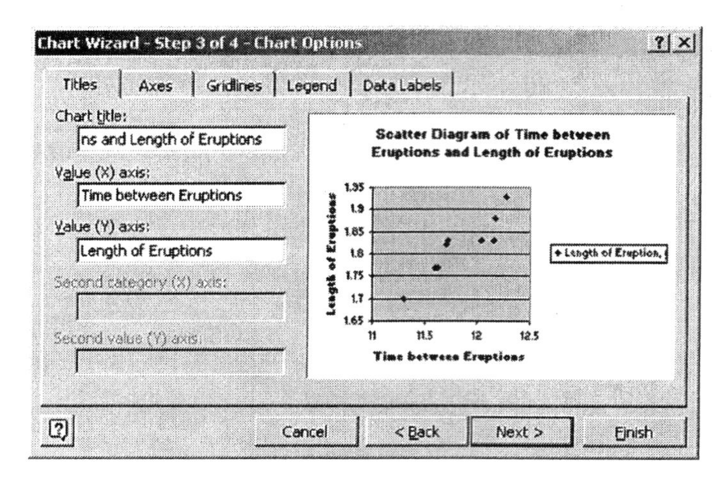

6. At the top of the Chart Options dialog box, click the **Legend** tab.

7. Click in the box to the left of **Show legend** to remove the checkmark. Click **Next>**.

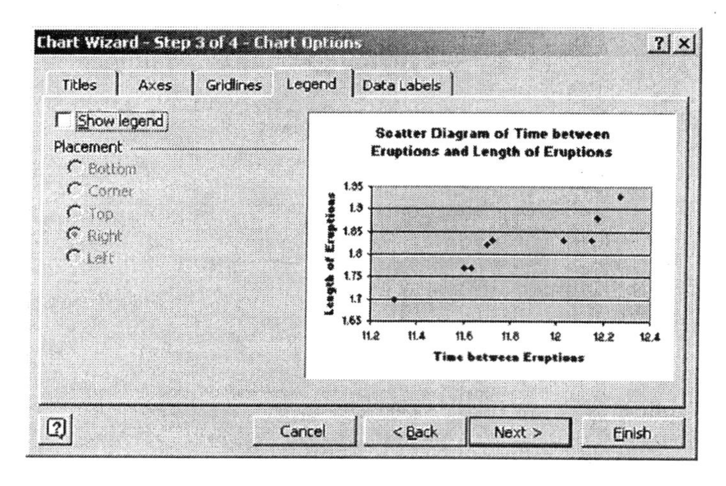

8. In the Chart Location dialog box, select **As new sheet**. Click **Finish**.

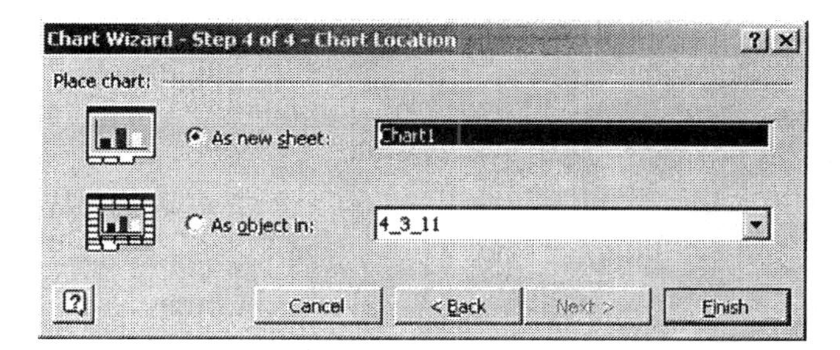

9. Place the regression line on the chart. To do this, **right click** directly on one of the dots. Then select **Add Trendline** from the shortcut menu that appears.

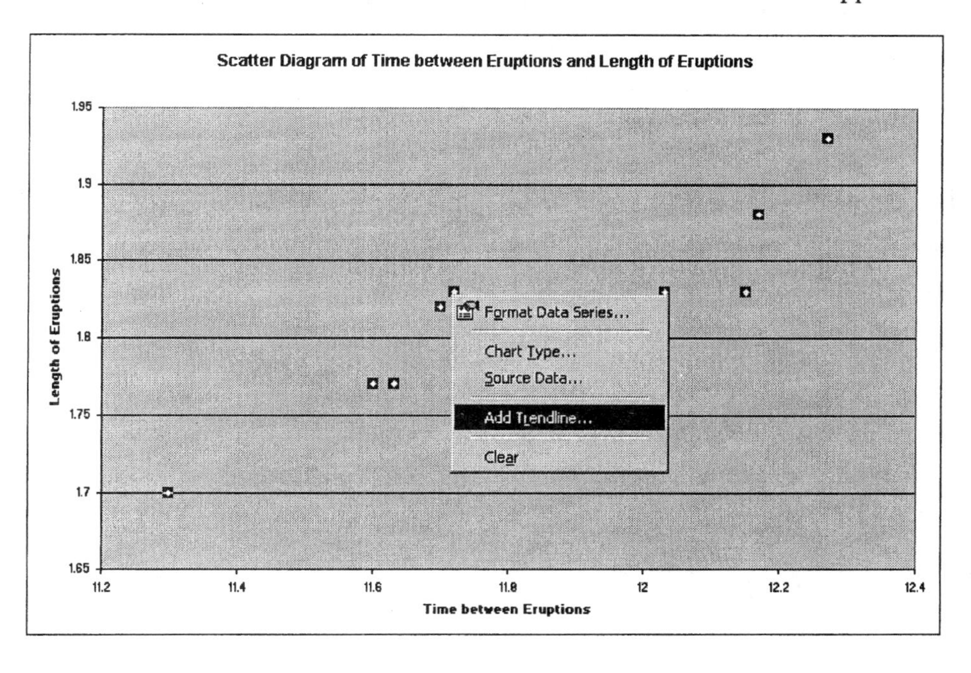

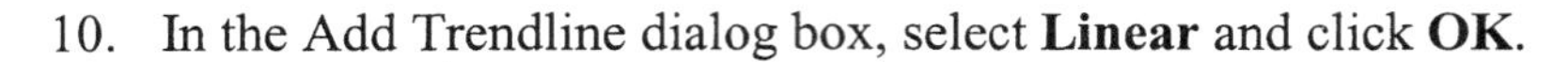

10. In the Add Trendline dialog box, select **Linear** and click **OK**.

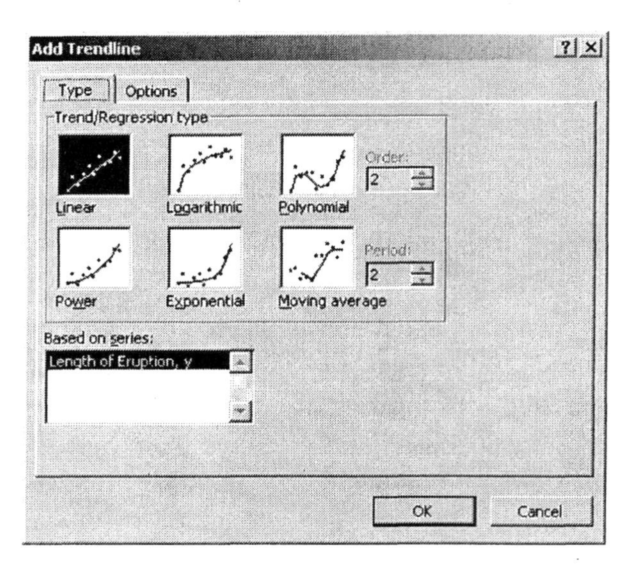

The completed scatter diagram should look similar to the diagram shown below.

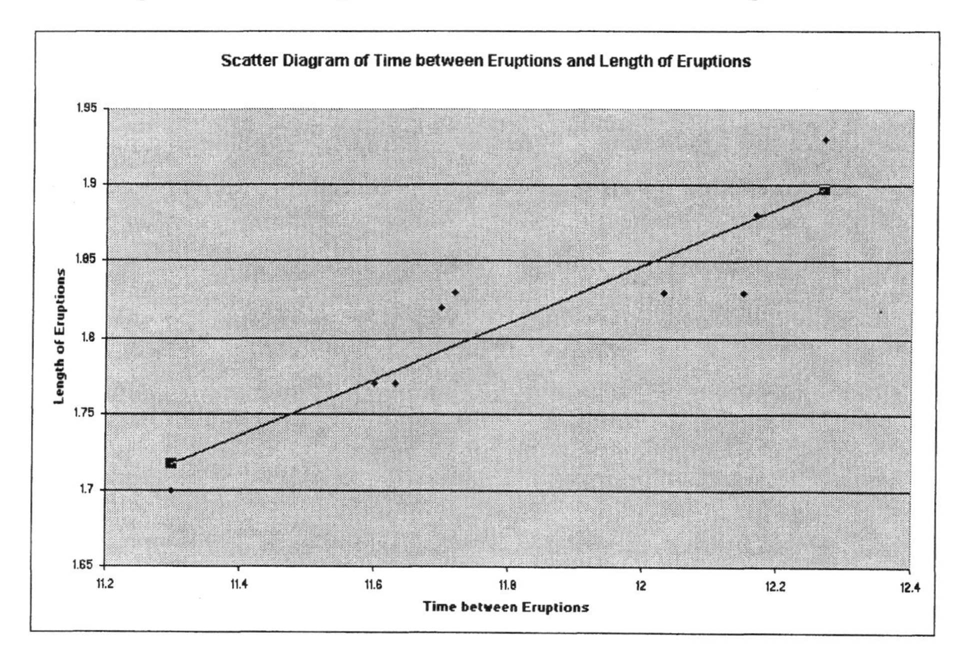

11. Next, you will carry out a regression analysis in order to obtain the residuals. Return to the worksheet that contains the data by clicking on the **4_3_11** sheet tab at the bottom of the screen.

12. At the top of the screen, click **Tools** and select **Data Analysis**.

If Data Analysis does not appear as a choice in the Tools menu, you will need to load the Microsoft Excel Analysis ToolPak add-in. Follow the procedure in Section GS 8.1 before continuing.

---

13. In the Data Analysis dialog box, select **Regression** and click **OK**.

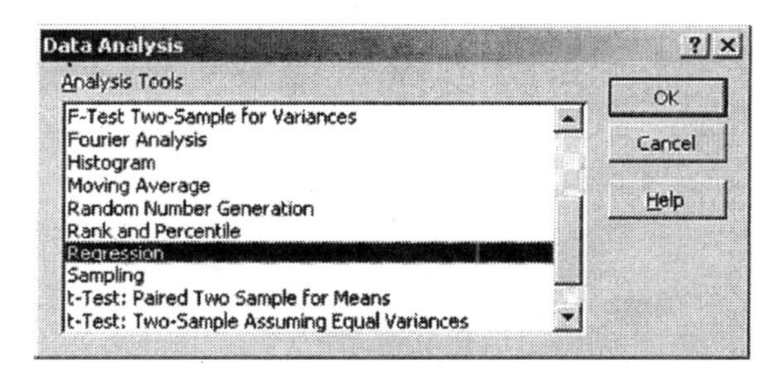

14. Complete the Regression dialog box as shown below. Be sure to select **Residuals** in the lower part of the dialog box. Click **OK**.

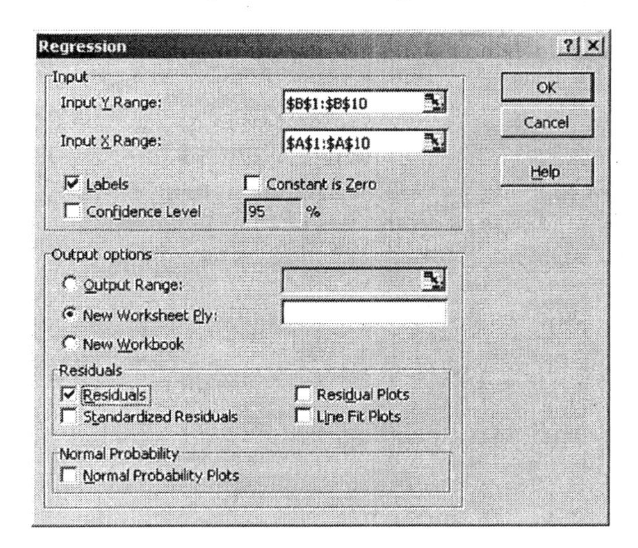

15. Next, you will copy the residuals to the worksheet containing the data so that you can use them in the residual plot. Click and drag over the residuals in the lower part

of the output so that the range **C24:C33** is highlighted. Then, at the top of the screen, click the copy button.

22	RESIDUAL OUTPUT		
23			
24	*Observatio*	*Length of E*	*Residuals*
25	1	1.878762	0.001238
26	2	1.778632	-0.00863
27	3	1.852803	-0.0228
28	4	1.875054	-0.04505
29	5	1.717442	-0.01744
30	6	1.791612	0.028388
31	7	1.897305	0.032695
32	8	1.773069	-0.00307
33	9	1.795321	0.034679
34			

16. Return to the worksheet containing the data by clicking on the **4_3_11** sheet tab at the bottom of the screen.

17. Click in cell **D1**and then, at the top of the screen, click the paste button.

	A	B	C	D	E	F	G
1	Time Betw	Length of Eruption, y		*Residuals*			
2	12.17	1.88		0.001238			
3	11.63	1.77		-0.00863			
4	12.03	1.83		-0.0228			
5	12.15	1.83		-0.04505			
6	11.3	1.7		-0.01744			
7	11.7	1.82		0.028388			
8	12.27	1.93		0.032695			
9	11.6	1.77		-0.00307			
10	11.72	1.83		0.034679			

18. To construct the residual plot, you will be using Time between Eruptions as the X variable and Residuals as the Y variable. These two variables need to be in adjacent columns of the worksheet. Copy the Time between Eruptions values to column C of the worksheet.

	A	B	C	D	E	F	G
1	Time Betw	Length of E	Time Betw	*Residuals*			
2	12.17	1.88	12.17	0.001238			
3	11.63	1.77	11.63	-0.00863			
4	12.03	1.83	12.03	-0.0228			
5	12.15	1.83	12.15	-0.04505			
6	11.3	1.7	11.3	-0.01744			
7	11.7	1.82	11.7	0.028388			
8	12.27	1.93	12.27	0.032695			
9	11.6	1.77	11.6	-0.00307			
10	11.72	1.83	11.72	0.034679			

19. Highlight the contents of both column C and column D by clicking and dragging over the range **C1:D10**. Then, at the top of the screen, click **Insert** and select **Chart**.

20. Under Chart type, select **XY (Scatter)**. Under Chart sub-type, select the topmost diagram. Click **Next>**.

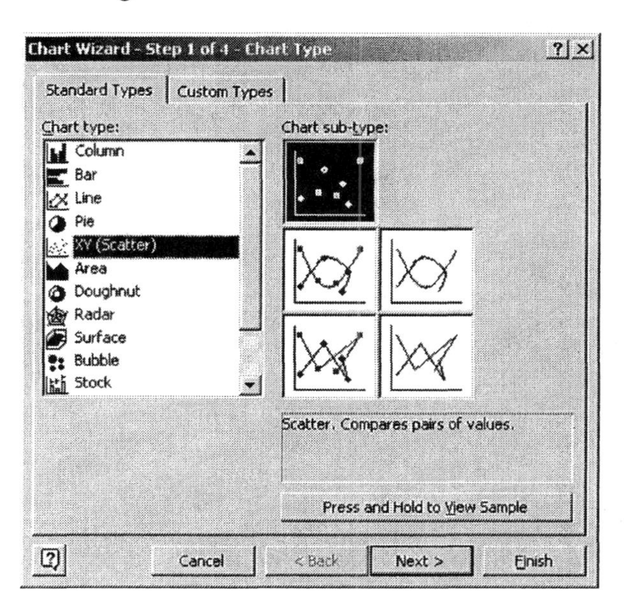

21. The data range should be the same as shown below. Make any necessary corrections. Then click **Next>**.

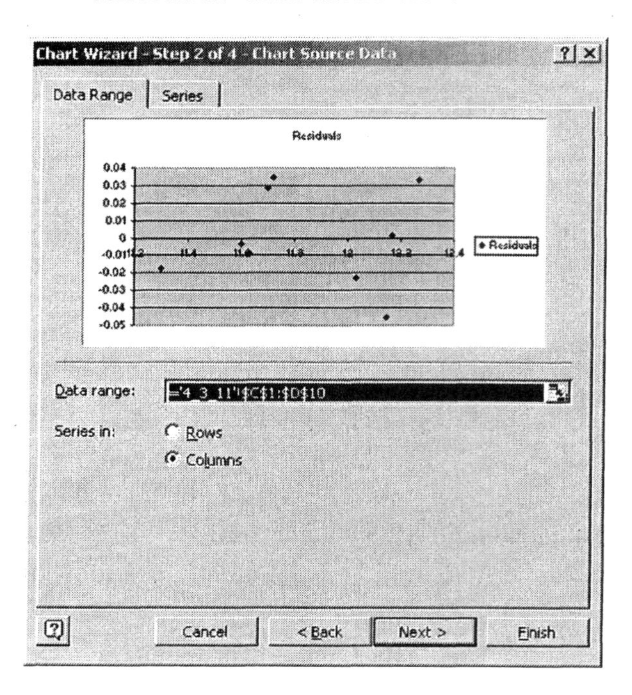

22. Click the **Titles** tab at the top of the Chart Options dialog box. Enter a title for the chart and labels for the X and Y axes. Chart title: **Residual Plot**. Value (X) axis: **Time between Eruptions**. Value (Y) axis: **Residuals.**

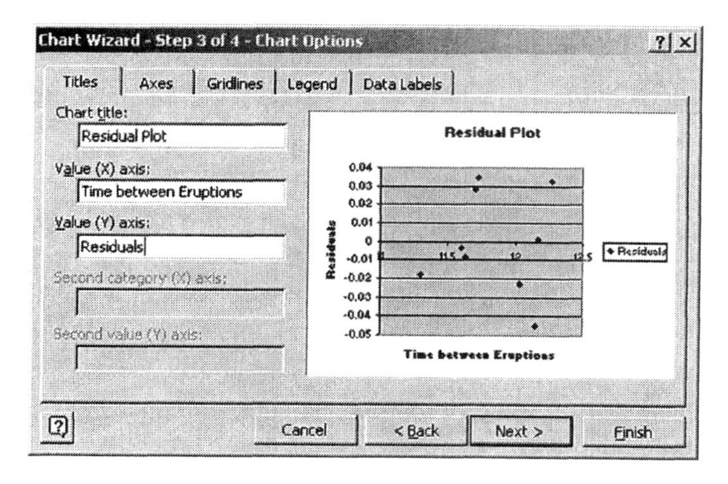

23. Click the **Legend** tab at the top of the Chart Options dialog box.

24. Click in the box to the left of **Show legend** to remove the checkmark. Click **Next>**.

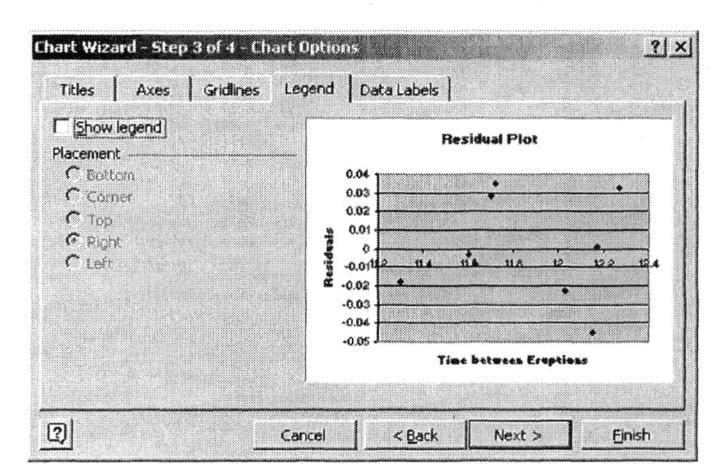

25. In the Chart Location dialog box, select **As new sheet**. Click **Finish**.

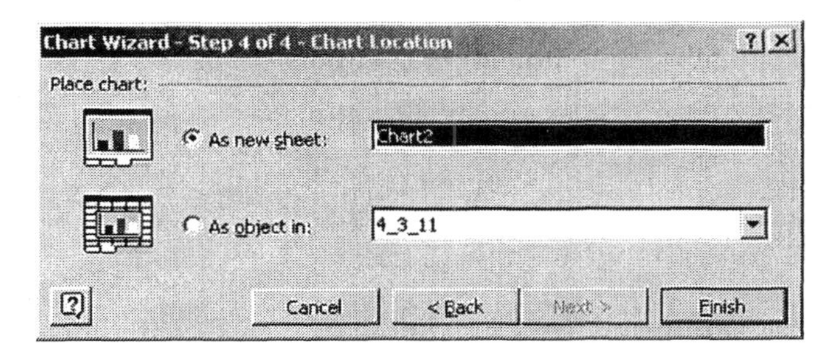

Your completed residual plot should look similar to the one shown below.

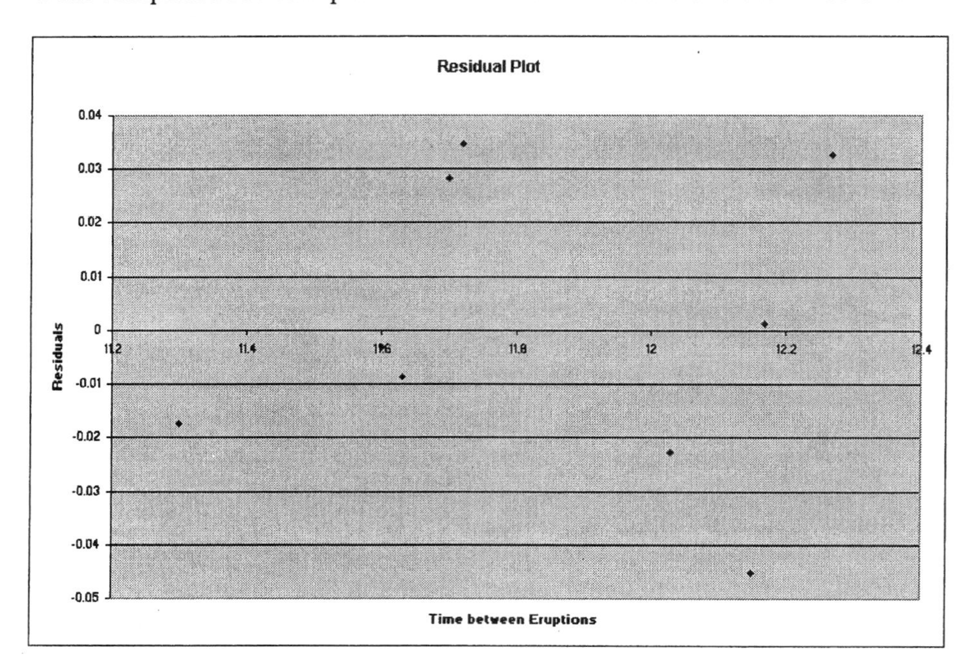

◀

Section 4.4 Nonlinear Regression: Transformations

▸ Example 4 (pg. 239)

Drawing a Scatter Diagram and Finding Regression Line of Transformed Data

1. Open worksheet "4_4_Ex4" in the Chapter 4 folder. The first few rows are shown below.

	A	B	C	D	E	F	G
1	Year, x	Closing Price, y					
2	1987 (x=1)	0.392					
3	1988 (x=2)	0.7652					
4	1989 (x=3)	1.1835					

2. Click in any cell of the data table. Then, at the top of the screen, click **Insert** and select **Chart**.

3. Under chart type, select **XY (Scatter)**. Under Chart sub-type, select the topmost diagram. Click **Next>**.

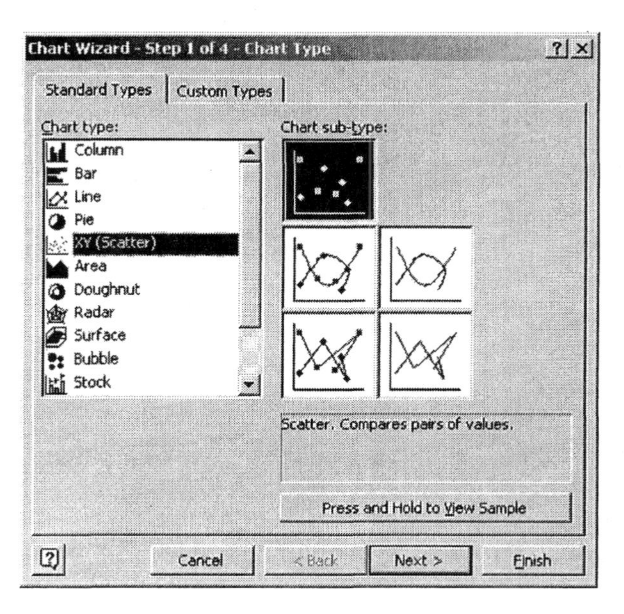

4. The data range should be the same as shown below. Make any necessary corrections. Then click **Next>**.

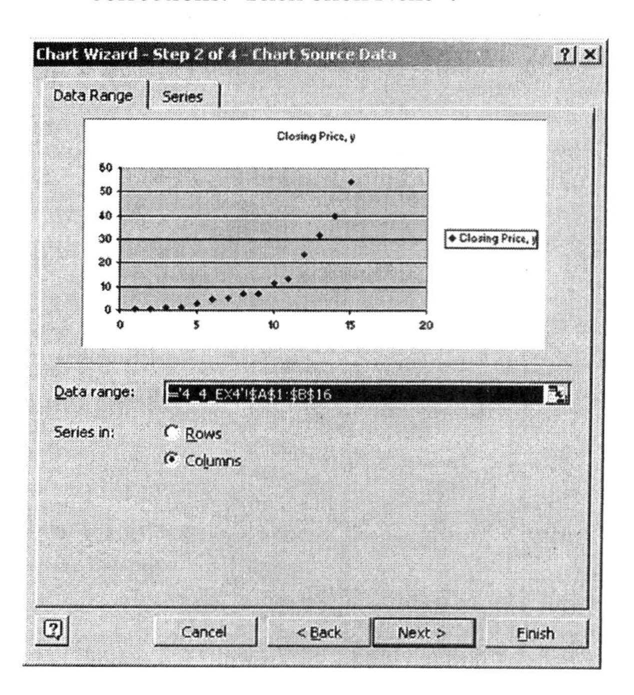

5. Enter a title for the chart and labels for the X and Y axes. Chart title: **Scatter Diagram of Year and Closing Price**. Value (X) axis: **Year**. Value (Y) axis: **Closing Price**.

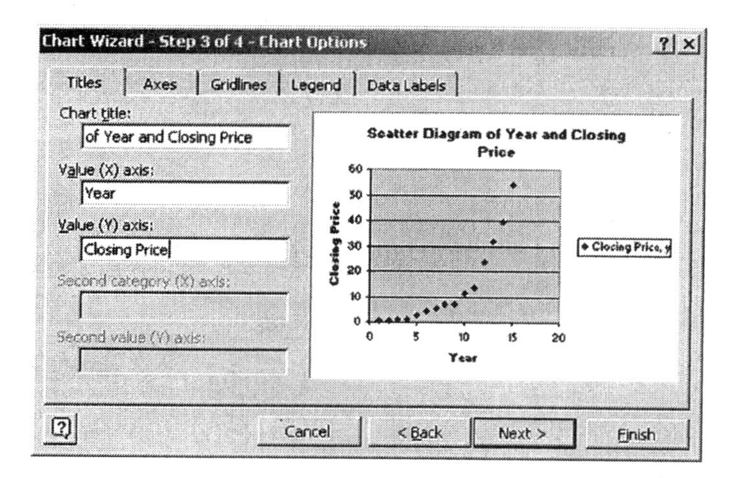

6. At the top of the Chart Options dialog box, click the **Legend** tab.

7. Click in the box to the left of **Show legend** to remove the checkmark. Click **Next>**.

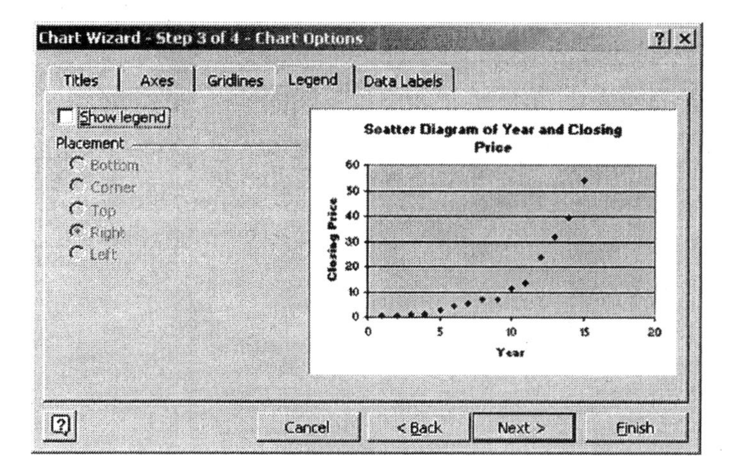

8. In the Chart Location dialog box, select **As new sheet**. Click **Finish**.

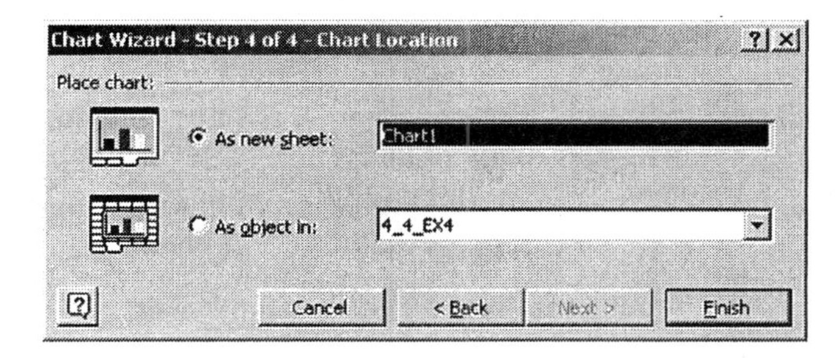

The completed scatter diagram should look similar to the one shown below.

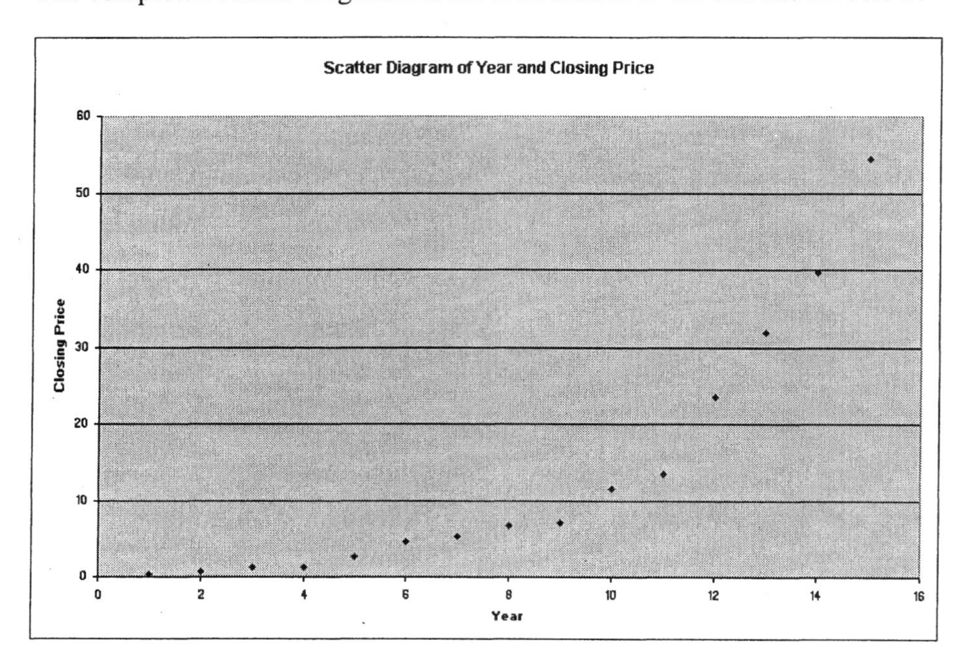

9. Next, you will determine the logarithm of the Y values. Return to the worksheet containing the data by clicking on the **4_4_EX4** sheet tab at the bottom of the screen. Enter the label **Y=log y** in cell D1.

	A	B	C	D	E	F	G
1	Year, x	Closing Price, y		Y=log y			
2	1987 (x = '	0.392					

10. Click in cell **D2** where you will place the log of 0.392.

11. At the top of the screen, click **Insert** and select **Function**.

12. Under Function category, select **Math & Trig**. Under Function name, select **LOG**. Click **OK**.

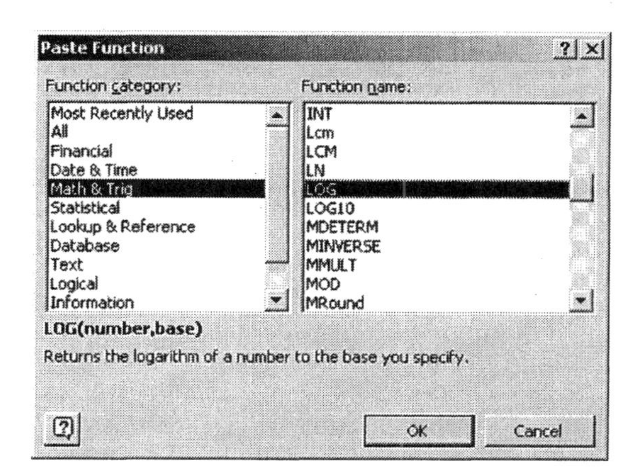

13. Complete the LOG dialog box as shown below. Rather than entering 0.392 in the Number window, you are entering the cell reference **B2**. By using the cell reference, you will be able to copy the LOG function rather than filling it out for each Y value. Click **OK**.

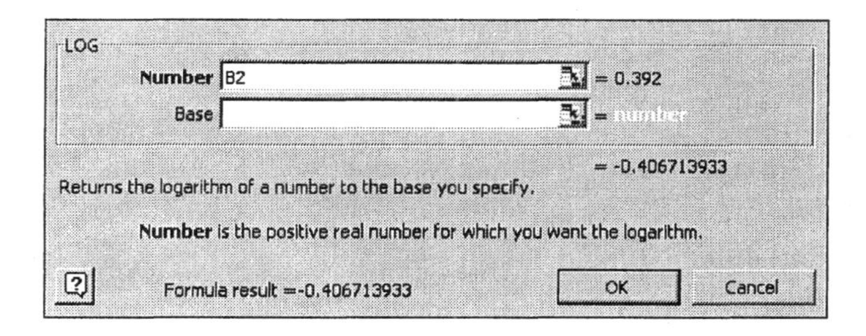

14. Copy the contents of cell D2 to cells D3 through D16.

	A	B	C	D	E	F	G
1	Year, x	Closing Price, y		Y=log y			
2	1987 (x =	0.392		-0.40671			
3	1988 (x =	0.7652		-0.11623			
4	1989 (x =	1.1835		0.073168			
5	1990 (x =	1.1609		0.064795			
6	1991 (x= 5	2.6988		0.431171			
7	1992 (x =	4.5381		0.656874			
8	1993 (x =	5.3379		0.72737			
9	1994 (x =	6.8032		0.832713			
10	1995 (x =	7.0328		0.847128			
11	1996 (x =	11.5585		1.062901			
12	1997 (x =	13.4799		1.129687			
13	1998 (x =	23.5424		1.371851			
14	1999 (x =	31.9342		1.504256			
15	2000 (x =	39.7277		1.599093			
16	2001 (x =	54.31		1.73488			

15. Next, you will need to enter the X values of 1, 2, 3, … 15 in a separate column of the worksheet. In their present form in column A, they will be treated as non-numeric data. Enter the label **X** in C1 and the values 1 through 15 in the cells immediately below.

	A	B	C	D	E	F	G
1	Year, x	Closing Pr	X	Y=log y			
2	1987 (x =	0.392	1	-0.40671			
3	1988 (x =	0.7652	2	-0.11623			
4	1989 (x =	1.1835	3	0.073168			
5	1990 (x =	1.1609	4	0.064795			
6	1991 (x= 5	2.6988	5	0.431171			
7	1992 (x =	4.5381	6	0.656874			
8	1993 (x =	5.3379	7	0.72737			
9	1994 (x =	6.8032	8	0.832713			
10	1995 (x =	7.0328	9	0.847128			
11	1996 (x =	11.5585	10	1.062901			
12	1997 (x =	13.4799	11	1.129687			
13	1998 (x =	23.5424	12	1.371851			
14	1999 (x =	31.9342	13	1.504256			
15	2000 (x =	39.7277	14	1.599093			
16	2001 (x =	54.31	15	1.73488			

16. At the top of the screen, click **Tools** and select **Data Analysis**.

If Data Analysis does not appear as a choice in the Tools menu, you will need to load the Microsoft Excel Analysis ToolPak add-in. Follow the procedure in Section GS 8.1 before continuing.

---

17. In the Data Analysis dialog box, select **Regression** and click **OK**.

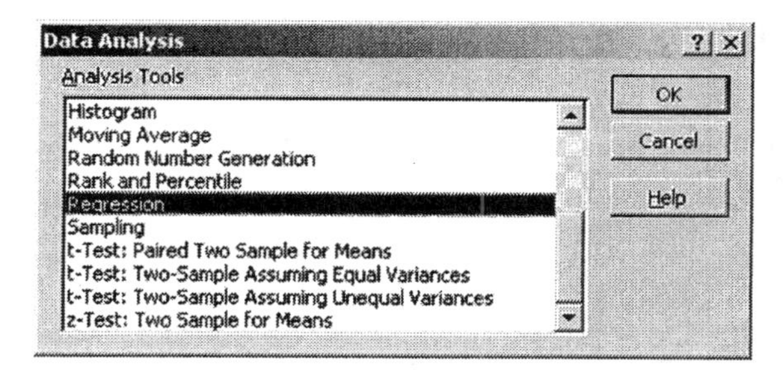

18. Complete the Regression dialog box as shown below. Click **OK**.

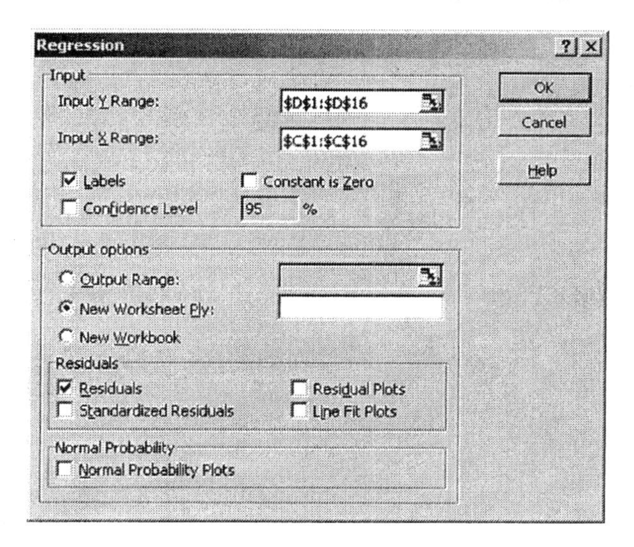

The output is displayed below. The intercept is –0.3952 and the slope is 0.1453.

	A	B	C	D	E	F	G
1	SUMMARY OUTPUT						
2							
3	*Regression Statistics*						
4	Multiple R	0.990603					
5	R Square	0.981294					
6	Adjusted F	0.979855					
7	Standard E	0.093126					
8	Observatio	15					
9							
10	ANOVA						
11		*df*	*SS*	*MS*	*F*	*ignificance F*	
12	Regressior	1	5.914318	5.914318	681.9605	1.28E-12	
13	Residual	13	0.112743	0.008673			
14	Total	14	6.027061				
15							
16		*Coefficient*	*tandard Err*	*t Stat*	*P-value*	*Lower 95%*	*Upper 95%*
17	Intercept	-0.39516	0.050601	-7.8093	2.91E-06	-0.50448	-0.28584
18	X	0.145336	0.005565	26.11437	1.28E-12	0.133313	0.157359

◀

Probability

CHAPTER 5

Section 5.1 Probability of Simple Events

▶ Example 6 (pg. 269) **Simulating Probabilities of Having a Baby Boy or a Baby Girl**

You will be generating random numbers selecting from 0 and 1 where 0 represents a boy and 1 represents a girl.

If the PHStat add-in has not been loaded, you will need to load it before continuing. Follow the instructions in Section GS 8.2

---

1. Open a new Excel worksheet.

2. Enter the numbers 0 and 1 in column A of the worksheet as shown below.

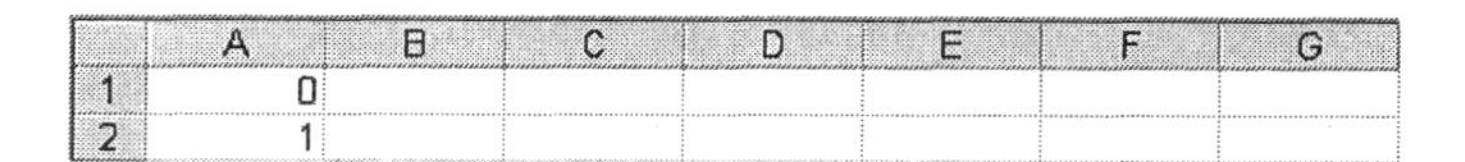

	A	B	C	D	E	F	G
1	0						
2	1						

3. At the top of the screen, click **Tools** and select **Data Analysis**.

4. Select **Sampling** and click **OK**.

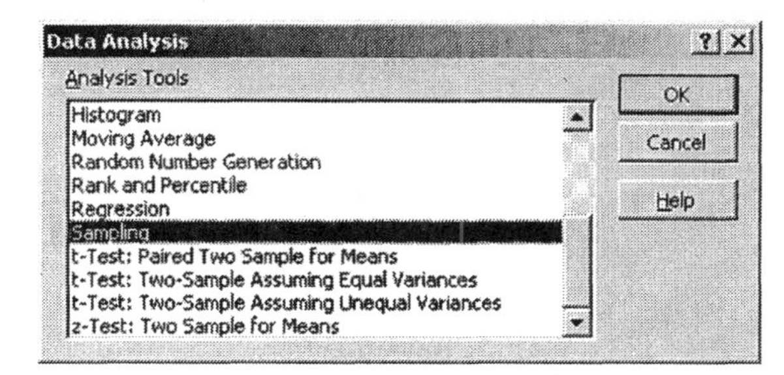

5. Complete the Sampling dialog box as shown below. You will be generating a sample of 100 numbers. The output will be placed in column B. Click **OK**.

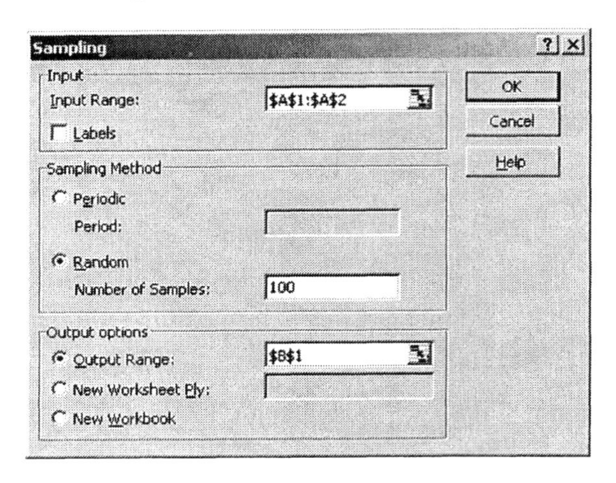

6. The first few rows of the output are displayed below. Because the numbers were generated randomly, it is not likely that your output will be exactly the same. Construct a frequency distribution to obtain a count of boys (0's) and girls (1's). At the top of the screen, click **PHStat** and select **Descriptive Statistics → One-Way Tables & Charts**.

	A	B	C	D	E	F	G
1	0	1					
2	1	1					
3		0					
4		0					

7. Complete the One-Way Tables & Charts dialog box as shown below. Click **OK**.

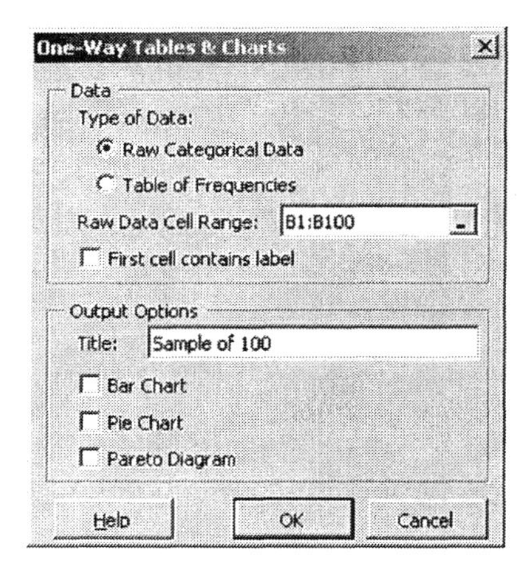

8. The frequency distribution table is displayed in a worksheet named "OneWayTable." In this example, you can see that there are 50 boys (0's) and 50 girls (1's).

	A	B	C	D	E	F	G
1	Sample of 100						
2							
3	Count of Variable						
4	Variable	Total					
5	0	50					
6	1	50					
7	Grand Total	100					

9. Generate a second sample with 1,000 random numbers. Go back to the worksheet containing the numbers 0 and 1 by clicking on the **Sheet1** tab near the bottom of the screen. Then click **Tools** and select **Data Analysis**. In the Data Analysis dialog box, select **Sampling** and click **OK**. Complete the Sampling dialog box as shown below. The 1,000 random numbers will be placed in column C. Click **OK**.

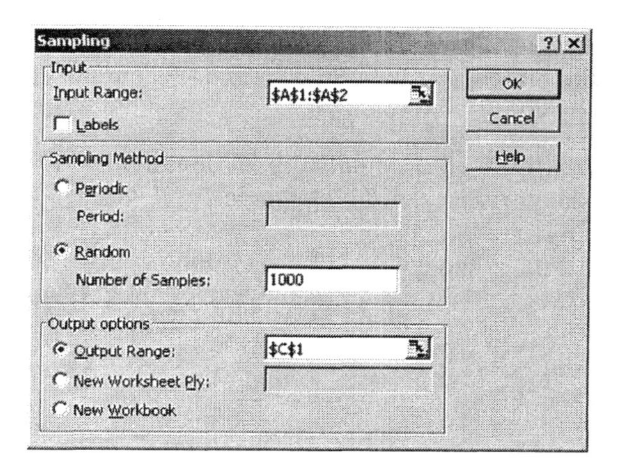

10. Construct a frequency distribution to obtain a count of the boys (0's) and girls (1's). At the top of the screen, click **PHStat** and select **Descriptive Statistics → One-Way Tables & Charts**.

11. Complete the One-Way Tables & Charts dialog box as shown below. Click **OK**.

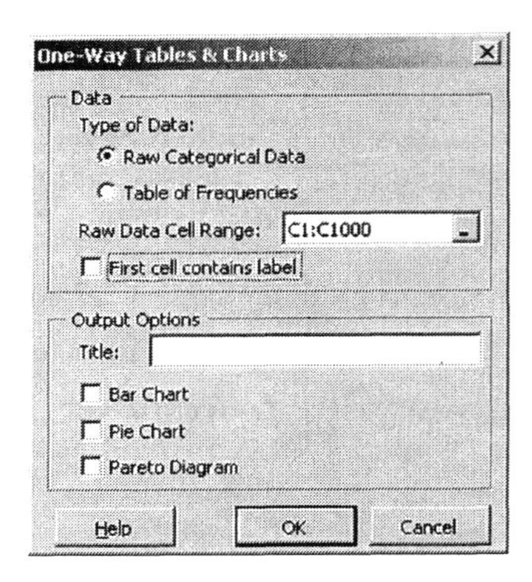

The frequency distribution is displayed below for the random sample that I generated. Your frequencies will likely be somewhat different. There are 509 boys and 491 girls in my sample.

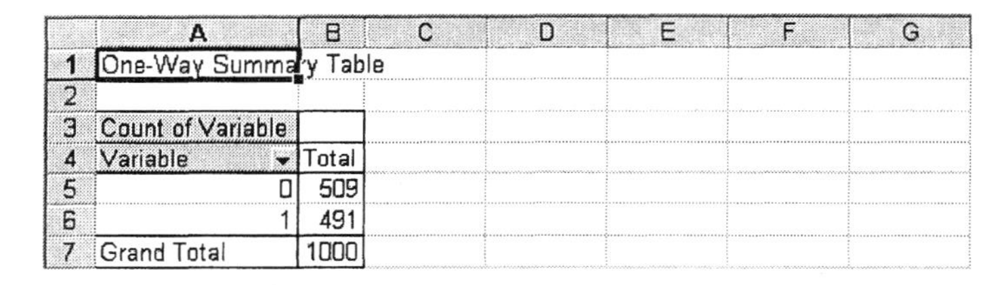

	A	B	C	D	E	F	G
1	One-Way Summary Table						
2							
3	Count of Variable						
4	Variable	Total					
5	0	509					
6	1	491					
7	Grand Total	1000					

◀

► Exercise 17 (pg. 272) Computing Probabilities

You will be computing probabilities associated with health-risk behaviors.

1. Open a new Excel worksheet.

2. Enter the information shown below. This is the college survey data displayed on page 272 of the textbook. You will be placing probabilities associated with the responses in column C.

	A	B	C	D	E	F	G
1	Response	Frequency	Probability				
2	Never	125					
3	Rarely	324					
4	Sometimes	552					
5	Most of the time	1257					
6	Always	2518					

3. Calculate the sum of the frequencies. To do this, first click in cell **B7** where you will place the sum. Then at the top of the screen, click the AutoSum button Σ .

	A	B	C	D	E	F	G
1	Response	Frequency	Probability				
2	Never	125					
3	Rarely	324					
4	Sometimes	552					
5	Most of the time	1257					
6	Always	2518					
7		=SUM(B2:B6)					

4. You should see =SUM(B2:B6). Make any necessary corrections, and then press [**Enter**]. Click in cell **C1** where you will place the probability associated with Never.

	A	B	C	D	E	F	G
1	Response	Frequency	Probability				
2	Never	125					
3	Rarely	324					
4	Sometimes	552					
5	Most of the time	1257					
6	Always	2518					
7		4776					

5. To compute the probability of Never, you will divide 125 by 4776. Carry out this division by entering a formula that uses cell references, **=B2/$B$7**, as shown below. The dollar signs are necessary to make the sum in B7 an absolute reference that will not change when you copy the formula. Press [**Enter**].

	A	B	C	D	E	F	G
1	Response	Frequency	Probability				
2	Never	125	=B2/$B$7				
3	Rarely	324					
4	Sometimes	552					
5	Most of the time	1257					
6	Always	2518					
7		4776					

6. Copy the contents of cell C2 to cells C3 through C7.

	A	B	C	D	E	F	G
1	Response	Frequency	Probability				
2	Never	125	0.026173				
3	Rarely	324	0.067839				
4	Sometimes	552	0.115578				
5	Most of the time	1257	0.263191				
6	Always	2518	0.527219				
7		4776	1				

Discrete Probability Distributions

CHAPTER 6

Section 6.1 Probability Distributions

▶ Exercise 15 (pg. 337)

Drawing a Probability Distribution, Computing the Mean and Variance

1. Open a new Excel worksheet.

2. Enter the information shown below. This is the data in Exercise 15 on page 337. The probabilities will be placed in column C. Enter the label **P(X=x)** in cell **C1**.

	A	B	C	D	E	F	G
1	x(age)	Frequency	P(X=x)				
2	5	2031					
3	6	2058					
4	7	2110					
5	8	2138					
6	9	2186					

3. Calculate the sum of the frequencies. To do this, first click in cell **B7** where you will place the sum. Then at the top of the screen, click the AutoSum button Σ.

	A	B	C	D	E	F	G
1	x(age)	Frequency	P(X=x)				
2	5	2031					
3	6	2058					
4	7	2110					
5	8	2138					
6	9	2186					
7		=SUM(B2:B6)					

4. You should see =SUM(B2:B6). Make any necessary corrections, and then press [**Enter**]. Click in cell **C2** where you will place the probability associated with x = 5.

	A	B	C	D	E	F	G
1	x(age)	Frequency	P(X=x)				
2	5	2031					
3	6	2056					
4	7	2110					
5	8	2138					
6	9	2186					
7		10523					

5. To compute the probability of x = 5, you will divide 2031 by 10523. Carry out this division by entering a formula that uses cell references, **=B2/$B$7**, as shown below. The dollar signs are necessary to make the sum in B7 an absolute reference that will not change when you copy the formula. Press [**Enter**].

	A	B	C	D	E	F	G
1	x(age)	Frequency	P(X=x)				
2	5	2031	=B2/$B$7				
3	6	2058					
4	7	2110					
5	8	2138					
6	9	2186					
7		10523					

6. Copy the contents of cell C2 to cells C3 through C7.

	A	B	C	D	E	F	G
1	x(age)	Frequency	P(X=x)				
2	5	2031	0.193006				
3	6	2058	0.195572				
4	7	2110	0.200513				
5	8	2138	0.203174				
6	9	2186	0.207735				
7		10523	1				

7. Highlight the range C1:C6. Then, at the top of the screen, click **Insert** and select **Chart**.

	A	B	C	D	E	F	G
1	x(age)	Frequency	P(X=x)				
2	5	2031	0.193006				
3	6	2058	0.195572				
4	7	2110	0.200513				
5	8	2138	0.203174				
6	9	2186	0.207735				
7		10523	1				

8. Under Chart type, select **Column**. Under Chart sub-type, select the leftmost diagram in the top row. Click **Next>**.

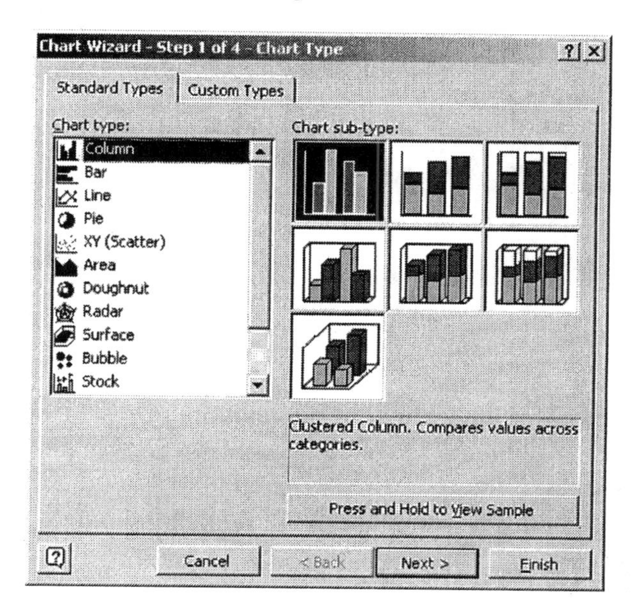

9. At the top of the Source Data dialog box, click the **Series** tab.

10. You want the values in column A to be the Category (X) axis labels. Click in the Category (X) axis labels window at the bottom of the dialog box. Then click and drag in the worksheet over the range A2:A6 to enter that information. Click **Next>**.

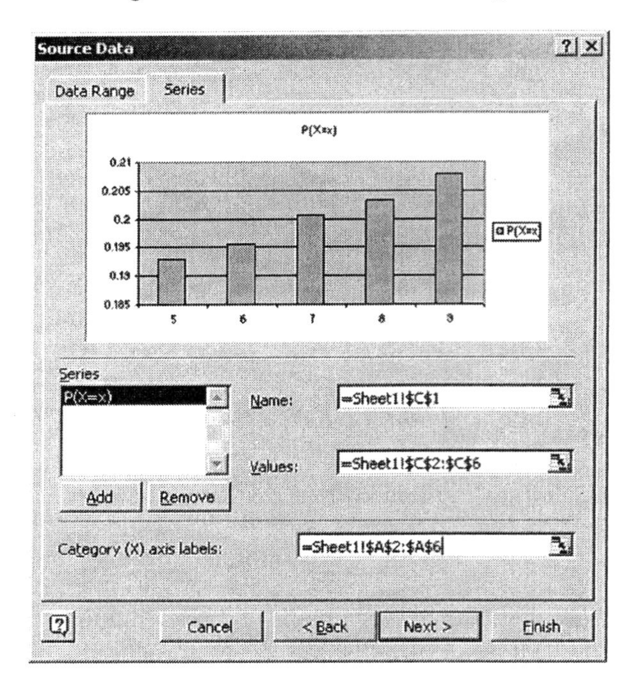

11. Enter a title for the chart and labels for the X and Y axes. Chart title: **Probability Distribution of 5-9 Year-old Males in the United States in 2000**. Category (X) axis: **Age**. Value (Y) axis: **Probability**.

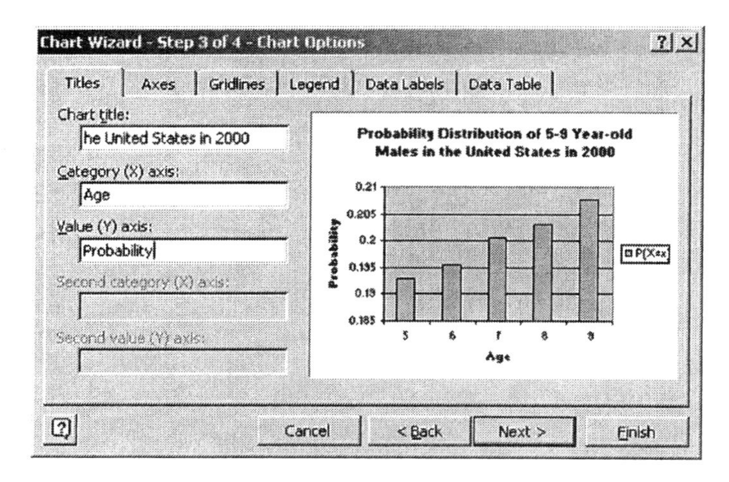

12. Click the **Legend** tab at the top of the Chart Options dialog box.

13. Click in the box to the left of **Show legend** to remove the checkmark. Click **Next>**.

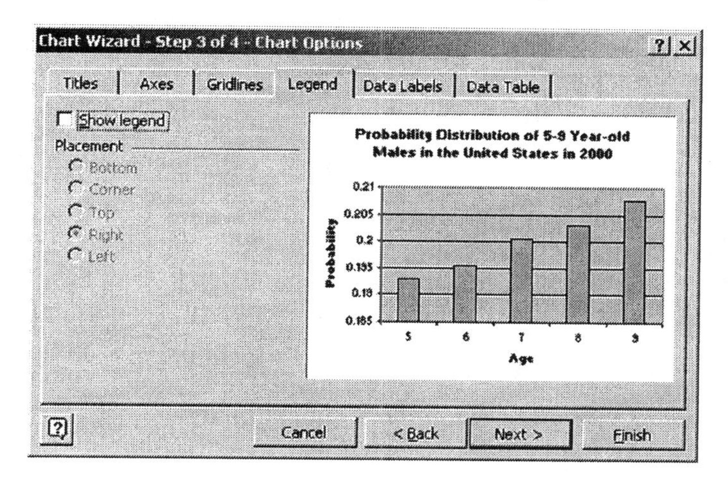

14. In the Chart Location dialog box, select **As new sheet**. Click **Finish**.

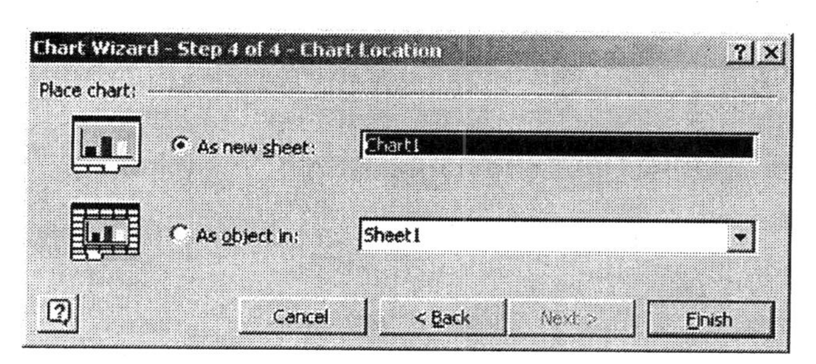

15. Remove the space between the vertical bars. **Right click** on one of the vertical bars. Select **Format Data Series** from the shortcut menu that appears.

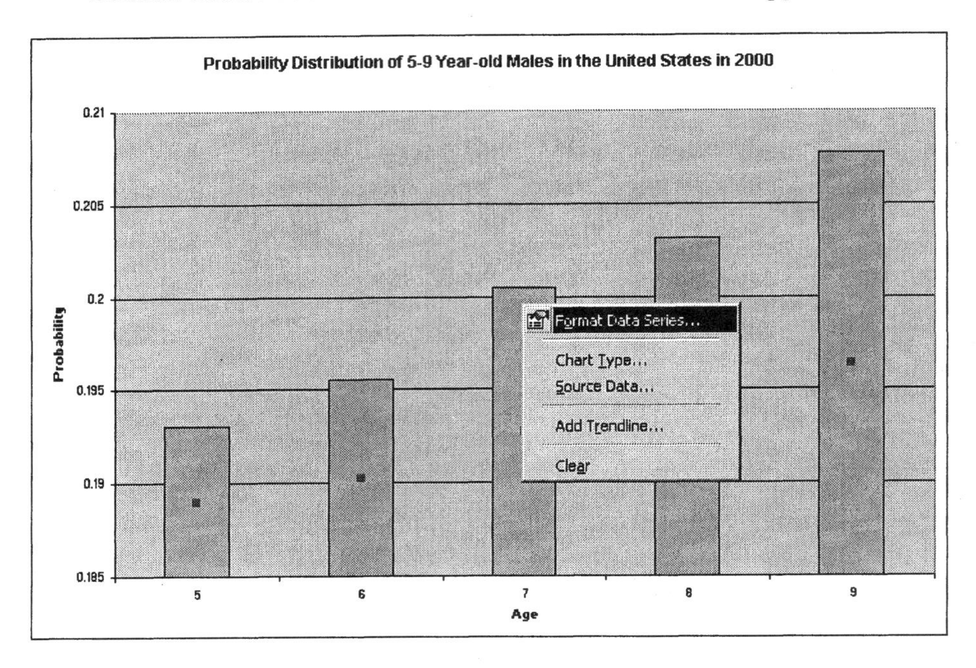

16. Click the **Options** tab at the top of the Format Data Series dialog box. Change the value in the Gap width box to **0**. Click **OK**.

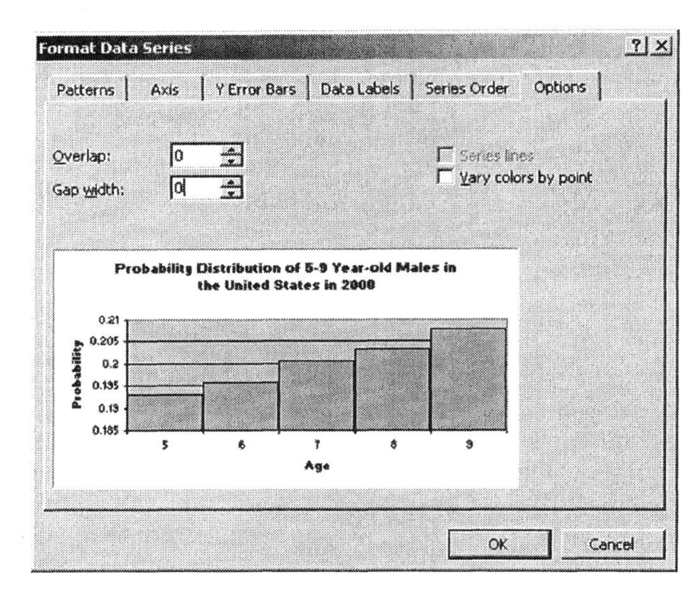

Your completed probability distribution should look similar to the one displayed below.

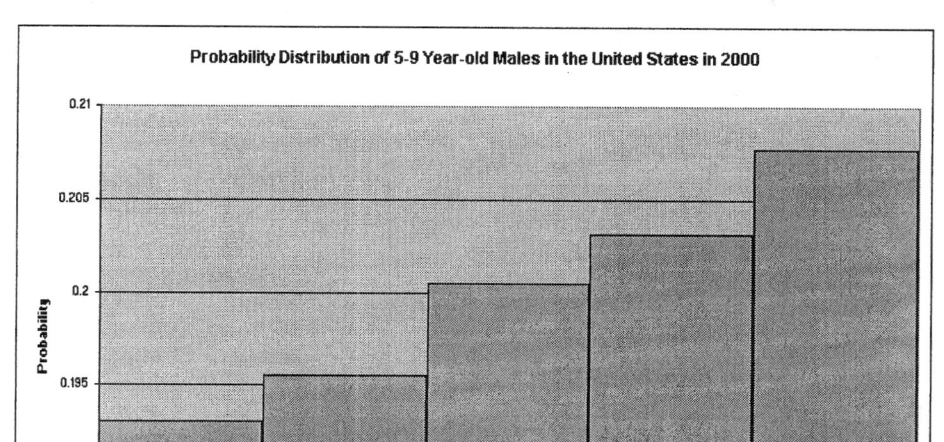

17. You will now compute the mean of this variable. Click the **Sheet1** tab at the bottom of the screen to return to the worksheet containing the data.

18. You will be applying the formula shown in the middle of page 330 in your textbook. Click in cell **D1** of the worksheet and enter the formula shown below, **=A2*C2**, to multiply the x value in column A by the probability in column C. Press [**Enter**].

	A	B	C	D	E	F	G
1	x(age)	Frequency	P(X=x)				
2	5	2031	0.193006	=A2*C2			
3	6	2058	0.195572				
4	7	2110	0.200513				
5	8	2138	0.203174				
6	9	2186	0.207735				
7		10523	1				

19. Copy the formula in cell D2 to cells D3 through D6.

	A	B	C	D	E	F	G
1	x(age)	Frequency	P(X=x)				
2	5	2031	0.193006	0.965029			
3	6	2058	0.195572	1.17343			
4	7	2110	0.200513	1.403592			
5	8	2138	0.203174	1.625392			
6	9	2186	0.207735	1.869619			
7		10523	1				

20. Add the products in column D. Click in cell **D7**. At the top of the screen, click the AutoSum button Σ. You should see =SUM(D2:D6). Press [**Enter**].

	A	B	C	D	E	F	G
1	x(age)	Frequency	P(X=x)				
2	5	2031	0.193006	0.965029			
3	6	2058	0.195572	1.17343			
4	7	2110	0.200513	1.403592			
5	8	2138	0.203174	1.625392			
6	9	2186	0.207735	1.869619			
7		10523	1	=SUM(D2:D6)			

The mean age is 7.0371 years.

	A	B	C	D	E	F	G
1	x(age)	Frequency	P(X=x)				
2	5	2031	0.193006	0.965029			
3	6	2058	0.195572	1.17343			
4	7	2110	0.200513	1.403592			
5	8	2138	0.203174	1.625392			
6	9	2186	0.207735	1.869619			
7		10523	1	7.037062			

21. You will now compute the variance of this variable. The formula is given on page 332 of your textbook. Click in cell **E2** of the worksheet. Enter the formula shown below, **=(A2-$D$7)^2*C2**, to multiply the squared deviation score by the probability. Note that the dollar signs are necessary for the cell reference of the mean. That reference must be an absolute reference that does not change when the formula is copied. Press [**Enter**].

	A	B	C	D	E	F	G
1	x(age)	Frequency	P(X=x)				
2	5	2031	0.193006	0.965029	=(A2-$D$7)^2*C2		
3	6	2058	0.195572	1.17343			
4	7	2110	0.200513	1.403592			
5	8	2138	0.203174	1.625392			
6	9	2186	0.207735	1.869619			
7		10523	1	7.037062			

22. Copy the contents of cell D2 to cells D3 through D6.

	A	B	C	D	E	F	G
1	x(age)	Frequency	P(X=x)				
2	5	2031	0.193006	0.965029	0.800901		
3	6	2058	0.195572	1.17343	0.210337		
4	7	2110	0.200513	1.403592	0.000275		
5	8	2138	0.203174	1.625392	0.188393		
6	9	2186	0.207735	1.869619	0.800431		
7		10523	1	7.037062			

23. Click in cell **E7** to place the sum there. At the top of the screen, click the AutoSum button Σ. In cell E7, you should see =SUM(E2:E6). Press [**Enter**].

	A	B	C	D	E	F	G
1	x(age)	Frequency	P(X=x)				
2	5	2031	0.193006	0.965029	0.800901		
3	6	2058	0.195572	1.17343	0.210337		
4	7	2110	0.200513	1.403592	0.000275		
5	8	2138	0.203174	1.625392	0.188393		
6	9	2186	0.207735	1.869619	0.800431		
7		10523	1	7.037062	=SUM(E2:E6)		

24. The variance of this variable is 2.0003. Click in cell **E8** to place the standard deviation there. Enter the formula shown below, **=SQRT(E7)**, to take the square root of the variance. Press [**Enter**].

	A	B	C	D	E	F	G
1	x(age)	Frequency	P(X=x)				
2	5	2031	0.193006	0.965029	0.800901		
3	6	2058	0.195572	1.17343	0.210337		
4	7	2110	0.200513	1.403592	0.000275		
5	8	2138	0.203174	1.625392	0.188393		
6	9	2186	0.207735	1.869619	0.800431		
7		10523	1	7.037062	2.000337		
8					=SQRT(E7)		

The standard deviation is 1.4143.

	A	B	C	D	E	F	G
1	x(age)	Frequency	P(X=x)				
2	5	2031	0.193006	0.965029	0.800901		
3	6	2058	0.195572	1.17343	0.210337		
4	7	2110	0.200513	1.403592	0.000275		
5	8	2138	0.203174	1.625392	0.188393		
6	9	2186	0.207735	1.869619	0.800431		
7		10523	1	7.037062	2.000337		
8					1.414333		

◀

▸ Exercise 25 (pg. 338) Simulating a Batter

1. Open a new Excel worksheet.

2. Enter the probability distribution from Problem 13, p. 336, as shown below.

	A	B	C	D	E	F	G
1	x	P(X=x)					
2	1	0.7					
3	2	0.21					
4	3	0.063					
5	4	0.0189					
6	5	0.006					
7	6	0.0021					

3. At the top of the screen, click **Tools** and select **Data Analysis**.

4. Select **Random Number Generation** and click **OK**.

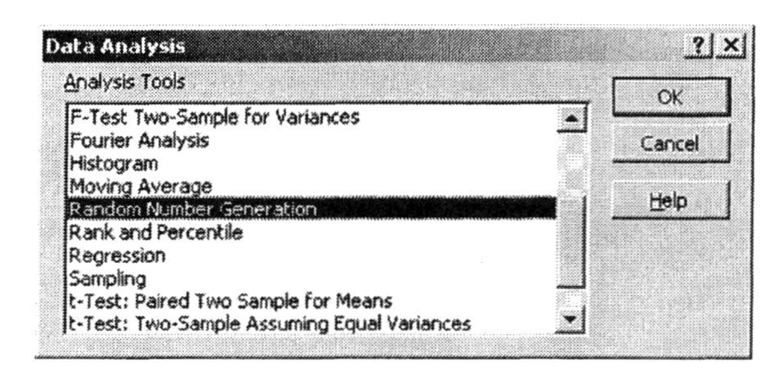

5. Complete the Random Number Generation dialog box as shown below.

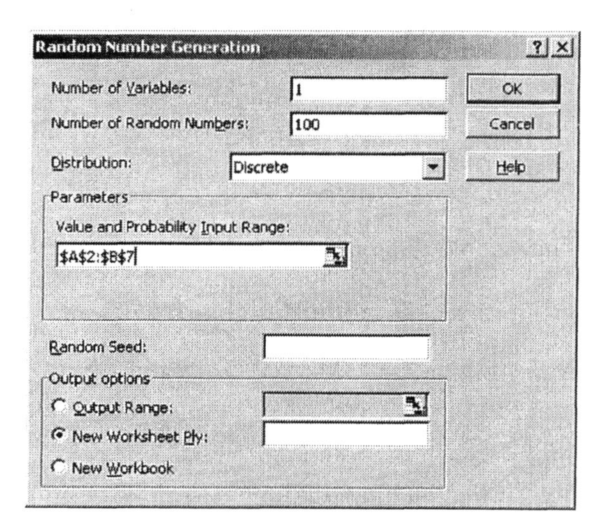

6. The first few lines of my output are displayed below. Because these numbers were generated randomly, it is not likely that your output will be exactly the same.

	A	B	C	D	E	F	G
1	1						
2	1						
3	4						
4	1						

7. To construct a frequency distribution using Excel's Pivot Table, you need a label in the topmost cell. Click in cell **A1**. At the top of the screen, click **Insert** and select **Rows**. Then type **X** in cell A1.

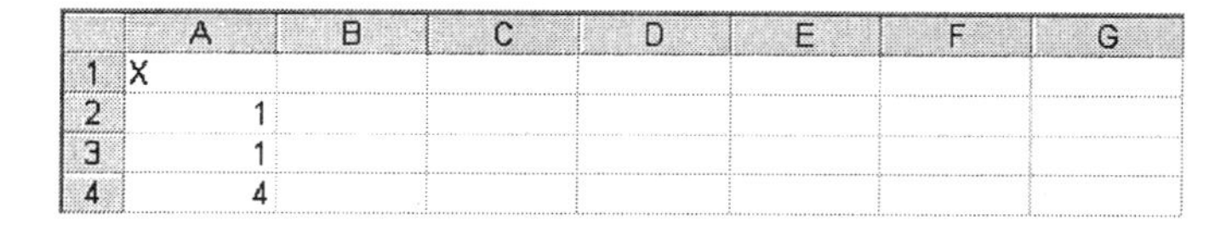

	A	B	C	D	E	F	G
1	X						
2	1						
3	1						
4	4						

8. You are now ready to construct a frequency distribution of these data. At the top of the screen, click **Data** and select **Pivot Table and Pivot Chart Report.**

9. Select **Microsoft Excel list** at the top of the dialog box. Select **Pivot Table** at the bottom of the dialog box. Click **Next>**.

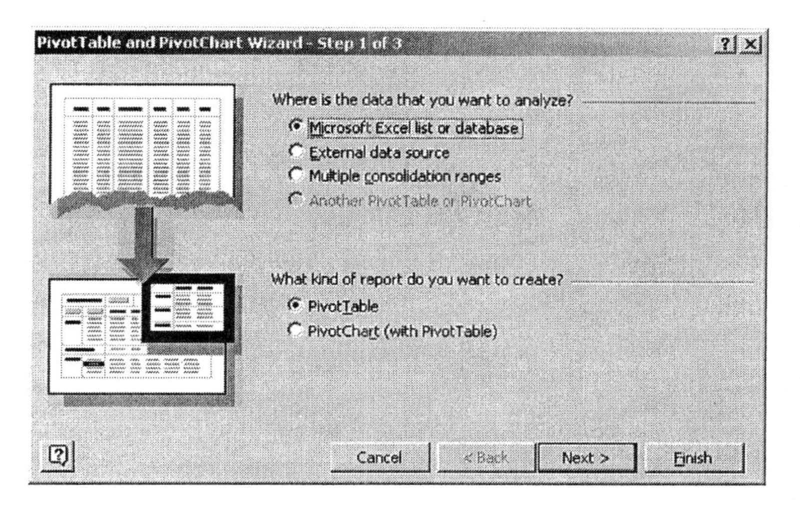

10. Enter the data range A1:A101 in the Step 2 dialog box as shown below. Click **Next>**.

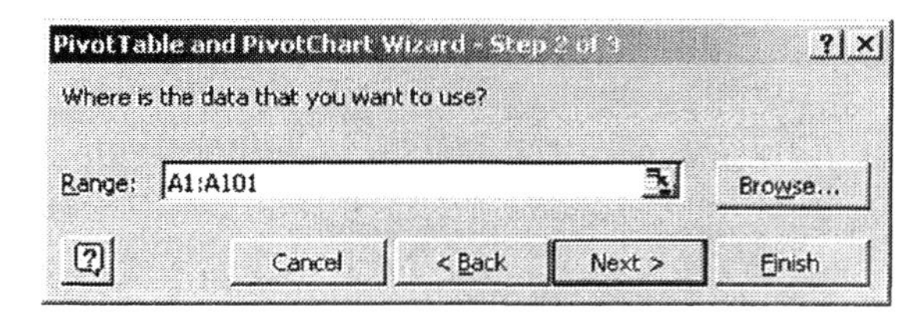

11. In the Step 3 dialog box, select **New worksheet**. At the bottom of the dialog box, click **Layout**.

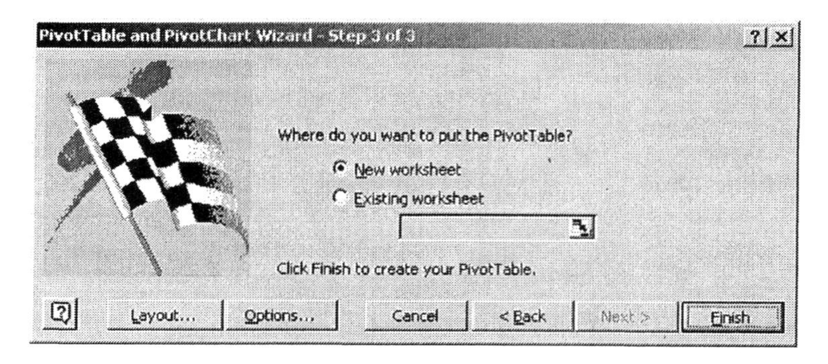

12. Drag the X field button at the right to ROW. Then drag the X field at the right to DATA.

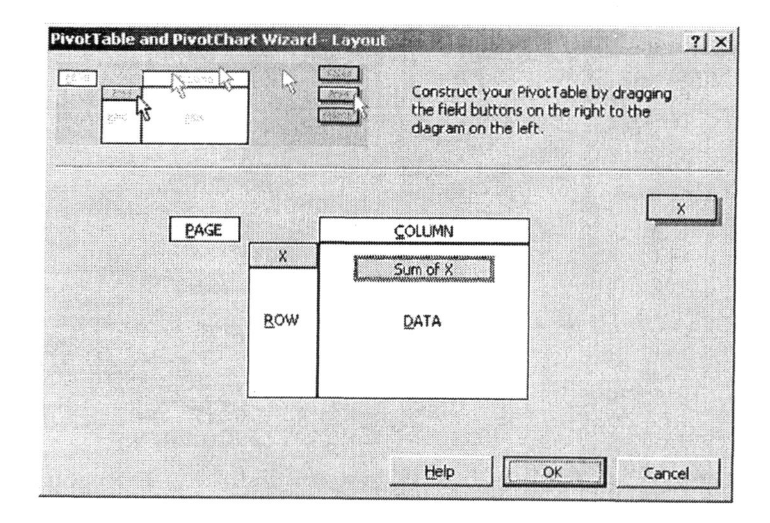

13. The default is Sum of X. You need to change this to count. Double click on the **Sum of X** button.

14. Under Summary by, select **Count**. Click **OK**. Also click **OK** in the Layout dialog box .

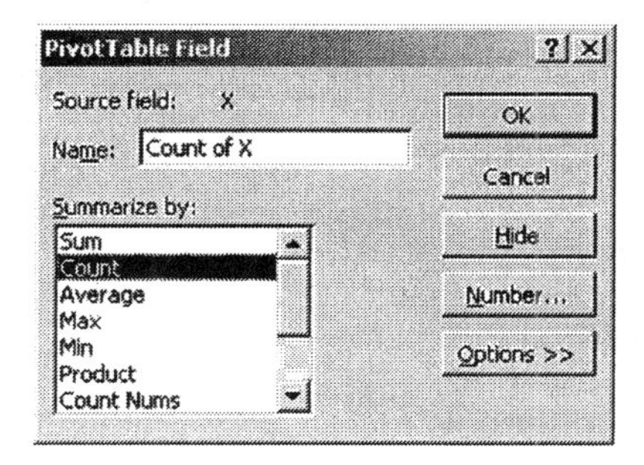

15. In the Step 3 dialog box, click **Finish**.

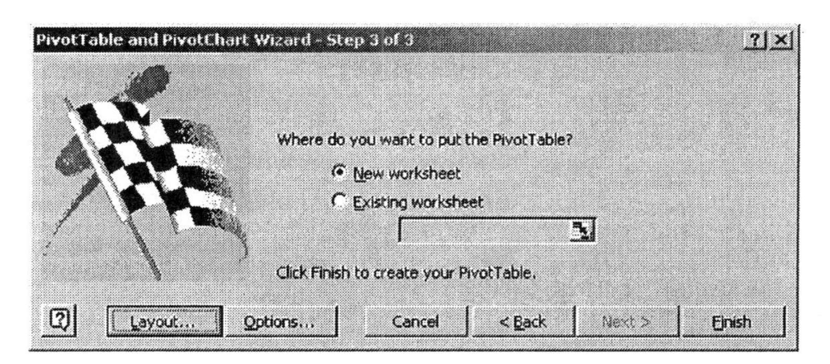

16. Enter the label **P(X=x)** in cell C4.

	A	B	C	D	E	F	G
1							
2							
3	Count of X						
4	X	Total	P(X=x)				
5	1	70					
6	2	19					
7	3	6					
8	4	4					
9	5	1					
10	Grand Total	100					

17. You will now compute the probability of observing x = 1 using a formula. Click in cell **C5**. Enter the formula **=B5/$B$10** as shown below. The dollar signs are necessary for the cell B10 address to make it an absolute reference that will not change when it is copied. Press [**Enter**].

	A	B	C	D	E	F	G
1							
2							
3	Count of X						
4	X	Total	P(X=x)				
5	1	70	=B5/$B$10				

18. Copy the contents of cell C5 to cells C6 through C10.

	A	B	C	D	E	F	G
1							
2							
3	Count of X						
4	X	Total	P(X=x)				
5	1	70	0.7				
6	2	19	0.19				
7	3	6	0.06				
8	4	4	0.04				
9	5	1	0.01				
10	Grand Total	100	1				

19. You will now compute the mean of this distribution. You will be applying the formula shown in the middle of page 330 in your textbook. Click in cell **D5** of the worksheet and enter the formula shown below to multiply the x value in column A by the probability in column C. Press [**Enter**].

	A	B	C	D	E	F	G
1							
2							
3	Count of X						
4	X	Total	P(X=x)				
5	1	70	0.7	=A5*C5			

20. Copy the formula in cell D5 to cells D6 through D9.

	A	B	C	D	E	F	G
1							
2							
3	Count of X						
4	X	Total	P(X=x)				
5	1	70	0.7	0.7			
6	2	19	0.19	0.38			
7	3	6	0.06	0.18			
8	4	4	0.04	0.16			
9	5	1	0.01	0.05			
10	Grand Total	100	1				

21. Add the products in column D. Click in cell **D10**. At the top of the screen, click the AutoSum button Σ. You should see =SUM(D5:D9). Press [**Enter**]. The mean is 1.47.

	A	B	C	D	E	F	G
1							
2							
3	Count of X						
4	X	Total	P(X=x)				
5	1	70	0.7	0.7			
6	2	19	0.19	0.38			
7	3	6	0.06	0.18			
8	4	4	0.04	0.16			
9	5	1	0.01	0.05			
10	Grand Total	100	1	=SUM(D5:D9)			

22. You will now compute the variance of this distribution. The formula is given on page 332 of your textbook. Click in cell **E5** of the worksheet. Enter the formula shown below to multiply the squared deviation score by the probability. Note that the dollar signs are necessary for the cell address of the mean. The address must be an absolute reference that does not change when the formula is copied. Press [**Enter**].

	A	B	C	D	E	F	G
1							
2							
3	Count of X						
4	X	Total	P(X=x)				
5	1	70	0.7	0.7	=(A5-$D$10)^2*C5		

23. Copy the contents of cell E5 to cells E5 through E9.

	A	B	C	D	E	F	G
1							
2							
3	Count of X						
4	X	Total	P(X=x)				
5	1	70	0.7	0.7	0.15463		
6	2	19	0.19	0.38	0.053371		
7	3	6	0.06	0.18	0.140454		
8	4	4	0.04	0.16	0.256036		
9	5	1	0.01	0.05	0.124609		
10	Grand Total	100	1	1.47			

24. Click in cell **E10** to place the sum there. At the top of the screen, click the AutoSum button Σ. In cell E10 you should see =SUM(E5:E9). Press [**Enter**]. The variance is 0.7291.

	A	B	C	D	E	F	G
1							
2							
3	Count of X						
4	X	Total	P(X=x)				
5	1	70	0.7	0.7	0.15463		
6	2	19	0.19	0.38	0.053371		
7	3	6	0.06	0.18	0.140454		
8	4	4	0.04	0.16	0.256036		
9	5	1	0.01	0.05	0.124609		
10	Grand Total	100	1	1.47	=SUM(E5:E9)		

25. Click in cell **E11** to place the standard deviation there. Enter the formula shown below to take the square root of the variance. Press [**Enter**].

	A	B	C	D	E	F	G
1							
2							
3	Count of X						
4	X	Total	P(X=x)				
5	1	70	0.7	0.7	0.15463		
6	2	19	0.19	0.38	0.053371		
7	3	6	0.06	0.18	0.140454		
8	4	4	0.04	0.16	0.256036		
9	5	1	0.01	0.05	0.124609		
10	Grand Total	100	1	1.47	0.7291		
11					=SQRT(E10)		

26. The standard deviation of this distribution is 0.8539. If you would like to repeat the simulation by performing 500 repetitions, start with step 1 of these instructions. In step 5, be sure to request 500 random numbers.

	A	B	C	D	E	F	G
1							
2							
3	Count of X						
4	X	Total	P(X=x)				
5	1	70	0.7	0.7	0.15463		
6	2	19	0.19	0.38	0.053371		
7	3	6	0.06	0.18	0.140454		
8	4	4	0.04	0.16	0.256036		
9	5	1	0.01	0.05	0.124609		
10	Grand Total	100	1	1.47	0.7291		
11					0.853874		

◀

Section 6.2 The Binomial Probability Distribution

▶ Example 5 (pg. 346) Constructing Binomial Probability Histograms

1. Open a new Excel worksheet and enter the information shown below. You will be using the BINOMDIST function to calculate binomial probabilities for x equal to 0 through 10.

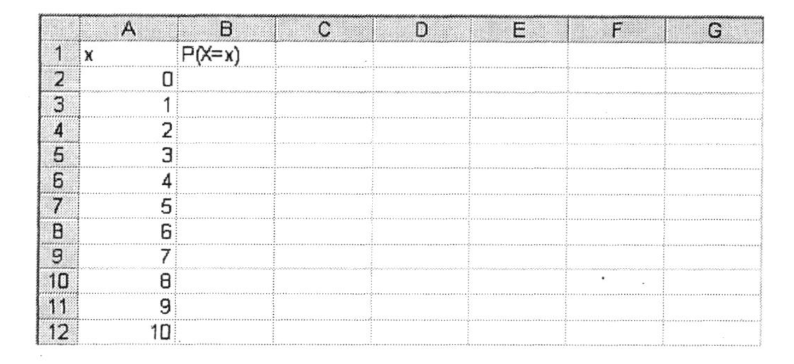

	A	B	C	D	E	F	G
1	x	P(X=x)					
2	0						
3	1						
4	2						
5	3						
6	4						
7	5						
8	6						
9	7						
10	8						
11	9						
12	10						

2. Click in cell **B2** below "P(X=x)." At the top of the screen, click **Insert** and select **Function**.

3. Under Function category, select **Statistical**. Under Function name, select **BINOMDIST**. Click **OK**.

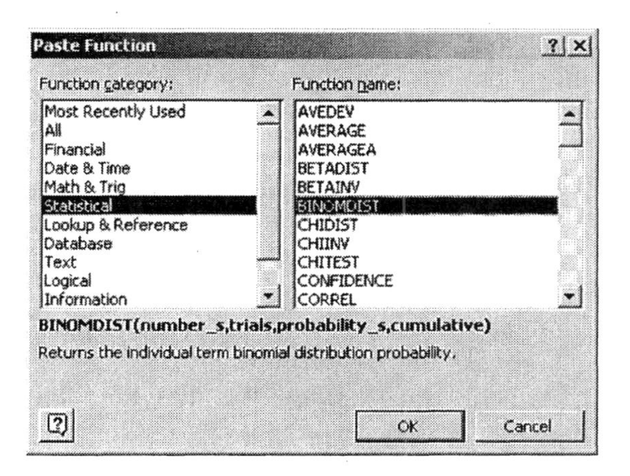

4. The instructions for the problem (a) tell you to use n = 10 and p = 0.2. Complete the BINOMDIST dialog box as shown below. Click **OK**.

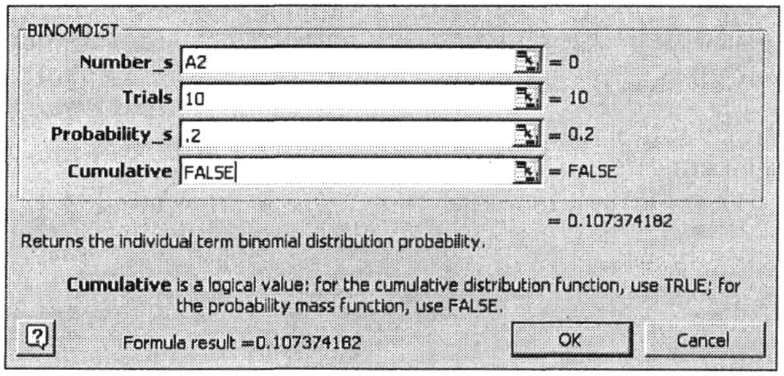

BINOMDIST

Number_s	A2	= 0
Trials	10	= 10
Probability_s	.2	= 0.2
Cumulative	FALSE	= FALSE

= 0.107374182

Returns the individual term binomial distribution probability.

Cumulative is a logical value: for the cumulative distribution function, use TRUE; for the probability mass function, use FALSE.

Formula result =0.107374182 OK Cancel

5. Copy the contents of cell B2 to cells B3 through B12.

	A	B	C	D	E	F	G
1	x	P(X=x)					
2	0	0.107374					
3	1	0.268435					
4	2	0.30199					
5	3	0.201327					
6	4	0.08808					
7	5	0.026424					
8	6	0.005505					
9	7	0.000786					
10	8	7.37E-05					
11	9	4.1E-06					
12	10	1.02E-07					

6. Click in any cell of the data table. Then, at the top of the screen, click **Insert** and select **Chart**.

7. Under Chart type, select **Column**. Under Chart sub-type, select the leftmost diagram in the first row. Click **Next>**.

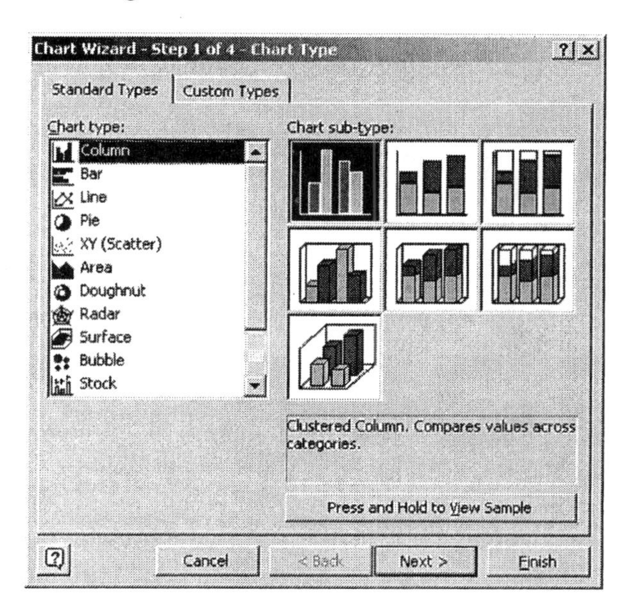

8. Check the accuracy of the data range in the Chart Source Data dialog box. It should read =Sheet1!$A$1:$B$12. Make any necessary corrections. Then click the **Series** tab at the top of the dialog box.

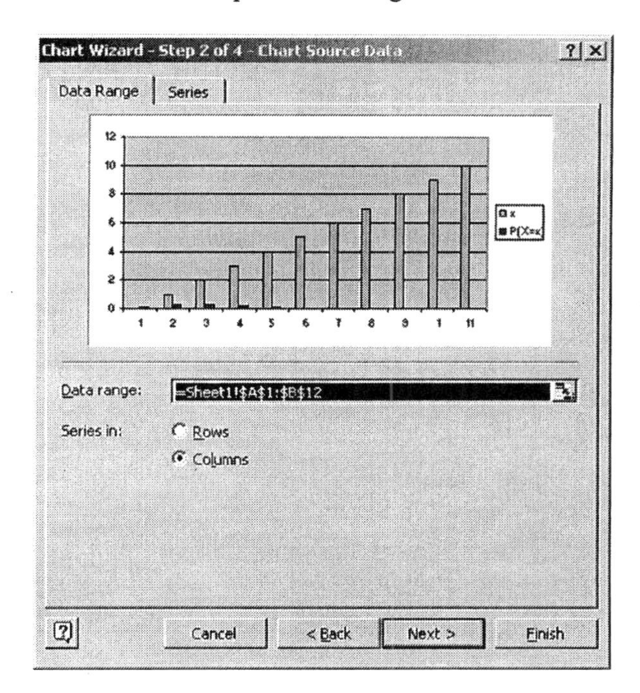

9. Under Series, select **X** and click the **Remove** button. Change the A's in the Values window entry to B's. Enter **=Sheet1!$A$2:$A$12** in the Category (X) axis labels window so that the values 0 through 10 will appear below the X axis. Click **Next>**.

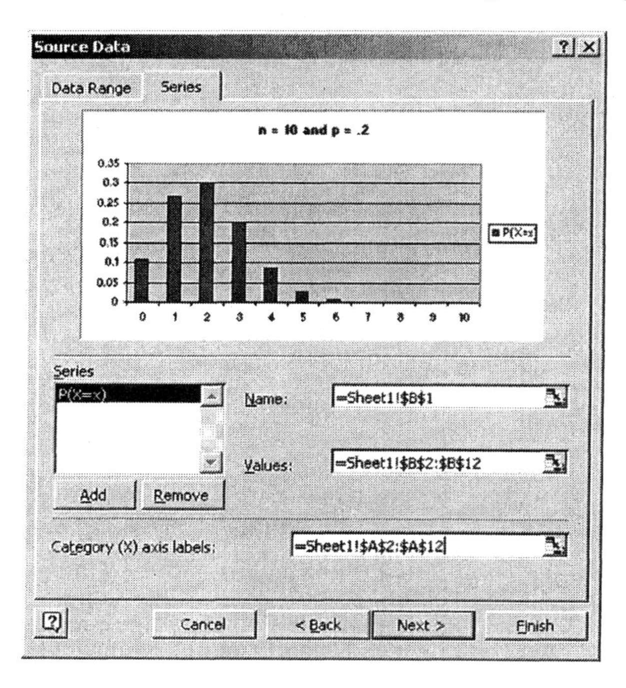

10. Click the **Titles** tab at the top. Enter a title for the chart and labels for the X and Y axes. Chart title: **n = 10 and p = .2**. Category (X) axis: **x**. Value (Y) axis: **Probability**. Click **Next>**.

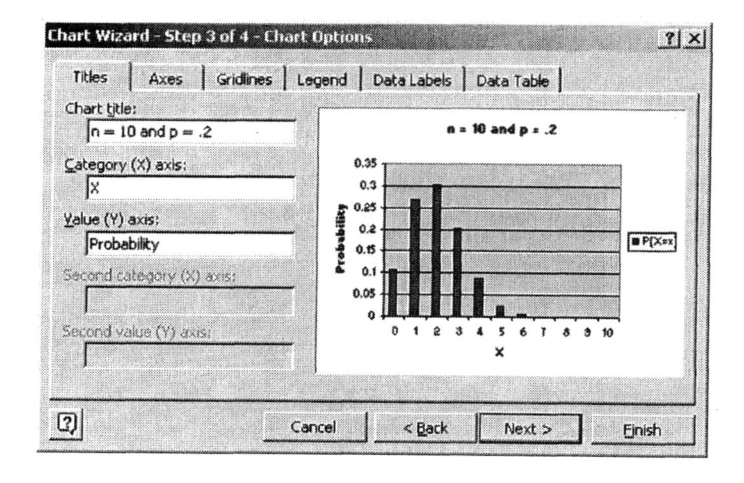

11. In the Chart Location dialog box, select **As new sheet**. Click **Finish**.

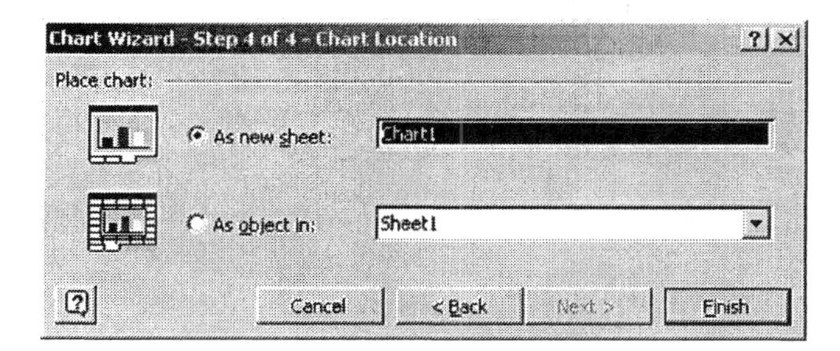

12. Next, remove the space between the vertical bars. **Right click** directly on one of the vertical bars and select **Format Data Series** from the shortcut menu that appears.

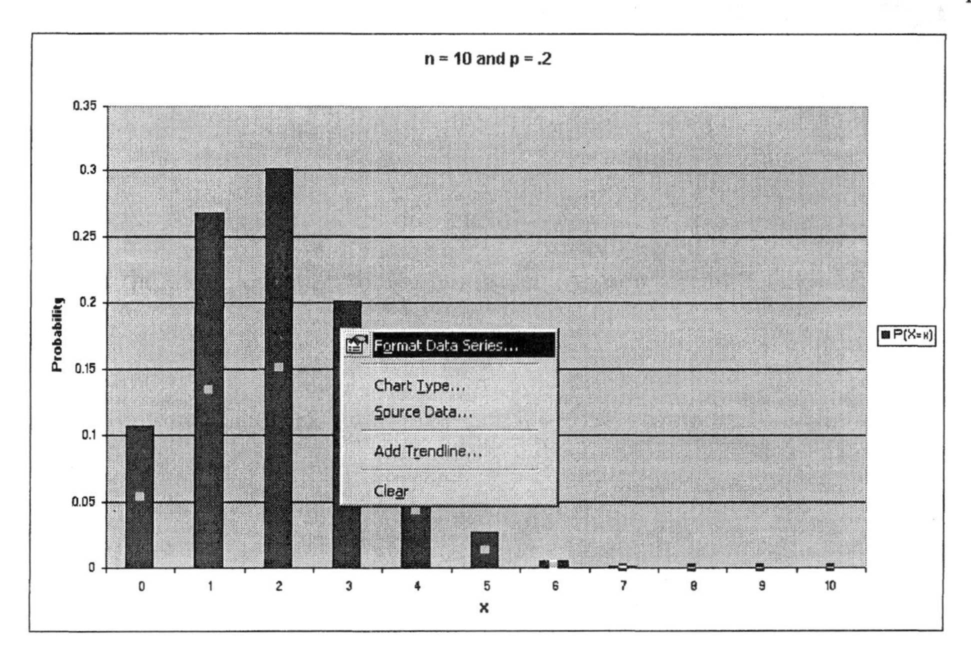

13. Click the **Options** tab at the top of the Format Data Series dialog box. Change the Gap width to **0**. Click **OK**. Your completed probability histogram should look similar to the one shown below in the dialog box.

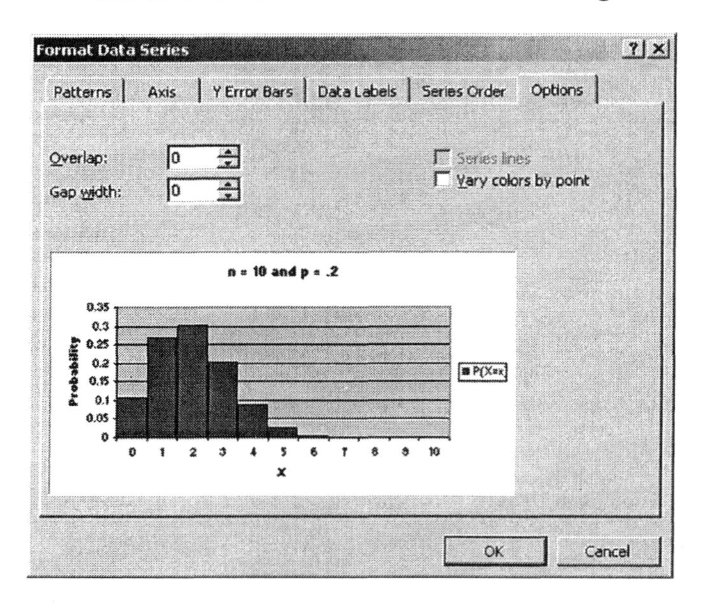

14. To construct the binomial probability histograms for problems (b) and (c) on page 346 of your textbook, repeat these instructions. Be sure to enter the appropriate probability in the BINOMDIST dialog box displayed in step 4.

◀

Section 6.3 The Poisson Probability Distribution

► Exercise 15 (pg. 367) — Constructing a Poisson Probability Distribution

1. Open a new Excel worksheet.

2. Enter the information shown below.

	A	B	C	D	E	F	G
1	x	P(X=x)					
2	0						
3	1						
4	2						
5	3						
6	4						
7	5						
8	6						
9	7						
10	8						
11	9						
12	10						
13	11						
14	12						
15	13						
16	14						
17	15						
18	16						

3. Click in cell **B2** where you will place the probability associated with x = 0. At the top of the screen, click **Insert** and select **Function.**

4. Under the Function category, select **Statistical**. Under Function name, select **POISSON**. Click **OK.**

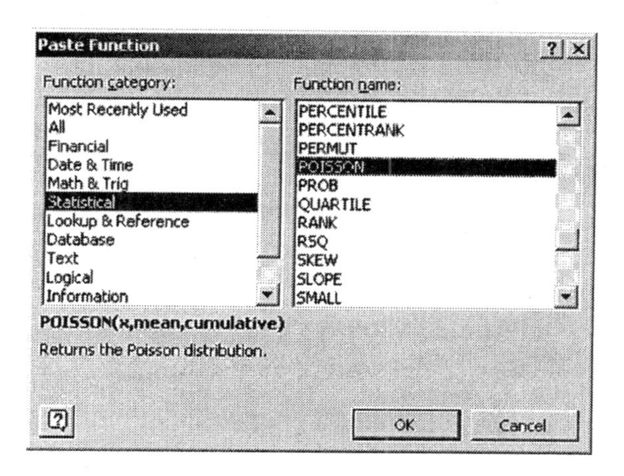

5. Complete the POISSON dialog box as shown below. The formula for the mean is shown at the bottom of page 362 in your textbook. The mean is equal to λt. For this exercise, λ = .2 and t = 30. So, λt = 6. Click **OK**.

POISSON

X: A2 = 0
Mean: 6 = 6
Cumulative: FALSE = FALSE

= 0.002478752

Returns the Poisson distribution.

Cumulative is a logical value: for the cumulative Poisson probability, use TRUE; for the Poisson probability mass function, use FALSE.

Formula result =0.002478752 | OK | Cancel

6. Copy the contents of cell B2 to cells B3 through B17. This is the probability distribution of x before the advertising.

	A	B	C	D	E	F	G
1	x	P(X=x)					
2	0	0.002479					
3	1	0.014873					
4	2	0.044618					
5	3	0.089235					
6	4	0.133853					
7	5	0.160623					
8	6	0.160623					
9	7	0.137677					
10	8	0.103258					
11	9	0.068838					
12	10	0.041303					
13	11	0.022529					
14	12	0.011264					
15	13	0.005199					
16	14	0.002228					
17	15	0.000891					
18	16	0.000334					

7. Open worksheet 6_3_15 in the chapter 6 folder. The first few rows are shown below.

	A	B	C	D	E	F	G
1	x (number	Frequency					
2	1	4					
3	2	5					
4	3	13					

8. Click in cell **B17** where you will place the sum of the frequencies. Then, at the top of the screen, click the AutoSum button Σ. You should see =SUM(B2:B17). Press [**Enter**]. The sum is 200.

16	15	0				
17	16	2				
18		=SUM(B2:B17)				

9. You will compute probabilities and place them in column C. Click in **C1** and enter the label **P(X=x)**.

	A	B	C	D	E	F	G
1	x (number	Frequency	P(X=x)				

10. Click in cell **C2** and enter the formula **=B2/$B$18** as shown below. The dollar signs are necessary for the B18 address because the sum must have an absolute reference that does not change when it is copied. Press [**Enter**].

	A	B	C	D	E	F	G
1	x (number	Frequency	P(X=x)				
2	1	4	=B2/$B$18				

11. Copy the contents of cell C2 to cells C3 through C17. This completes the probability distribution of x after the advertising.

	A	B	C	D	E	F	G
1	x (number	Frequency	P(X=x)				
2	1	4	0.02				
3	2	5	0.025				
4	3	13	0.065				
5	4	23	0.115				
6	5	25	0.125				
7	6	28	0.14				
8	7	25	0.125				
9	8	27	0.135				
10	9	21	0.105				
11	10	15	0.075				
12	11	5	0.025				
13	12	3	0.015				
14	13	2	0.01				
15	14	2	0.01				
16	15	0	0				
17	16	2	0.01				
18		200	1				

The Normal Probability Distribution

CHAPTER 7

Section 7.1 Properties of the Normal Distribution

► Exercise 23 (pg. 391) | Constructing a Relative Frequency Histogram

1. Open worksheet "7_1_23" in the Chapter 7 folder. The first few rows are shown below.

	A	B	C	D	E	F	G
1	Pitching Wedge Distance						
2	100						
3	104						
4	104						

2. Click in any cell of column A that has an entry and then, at the top of the screen, click **Data** and select **Pivot Table and Pivot Chart Report**.

3. At the top of the Step 1 dialog box, select **Microsoft Excel list**. At the bottom of the dialog box, select **Pivot Table**. Click **Next>**.

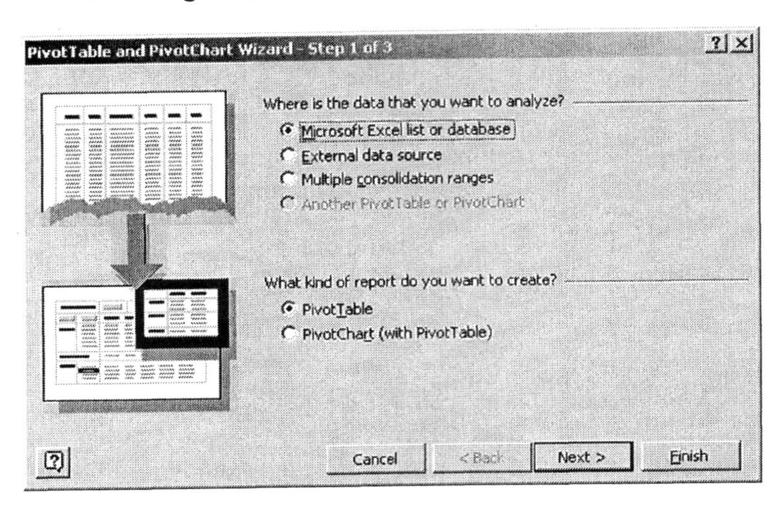

4. The data range $A$1:$A$76 should automatically appear in the Range window. Make any necessary corrections. Then click **Next>**.

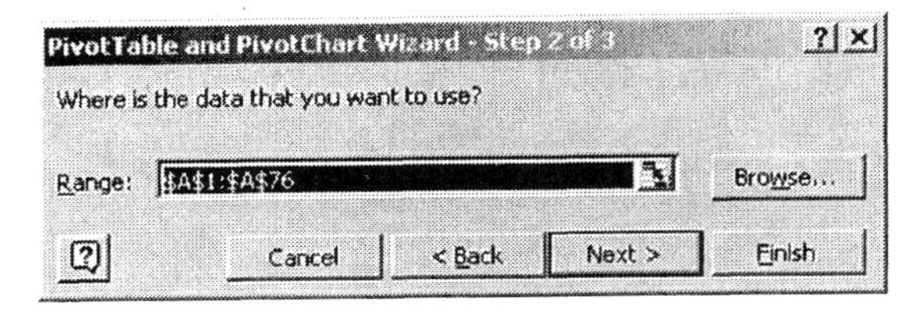

5. Select **New worksheet**. Click the **Layout** button at the bottom of the dialog box.

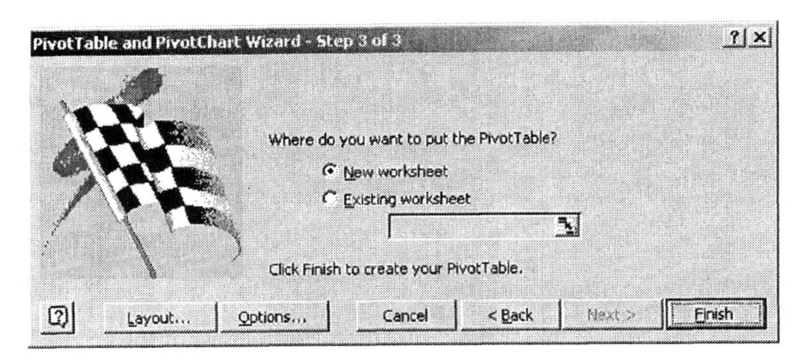

6. Drag the Pitching field button to ROW, and drag the Pitching field button to DATA.

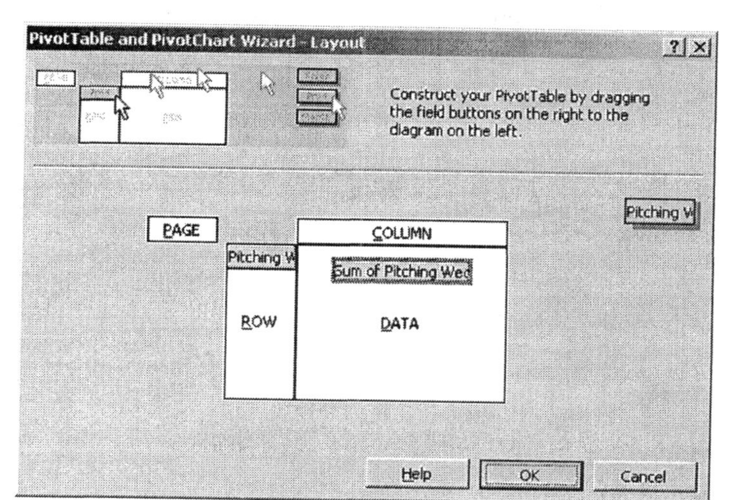

7. The default summary measure is Sum. You want Count. To make this change, double click on the **Sum of Pitching** button.

8. Under Summarize by, select **Count**. Click **OK**. Also click **OK** in the Layout dialog box. Click **Finish** in the Step 3 dialog box.

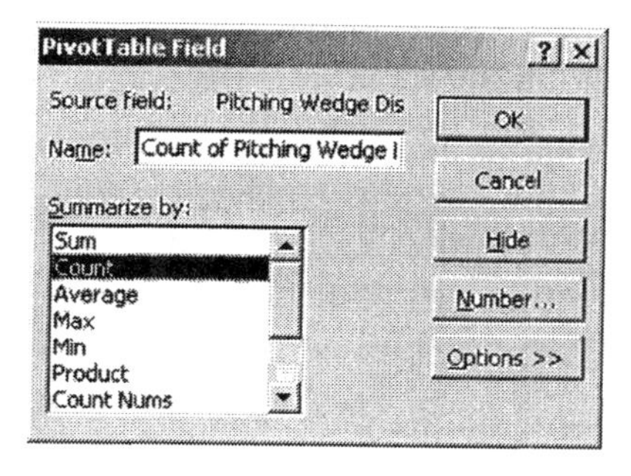

9. The output consists of frequencies and you want relative frequencies. You will need to compute the relative frequencies. Enter the label **Relative Frequency** in cell C4. Then click in cell **C5** where you will place the relative frequency associated with a distance of 94.

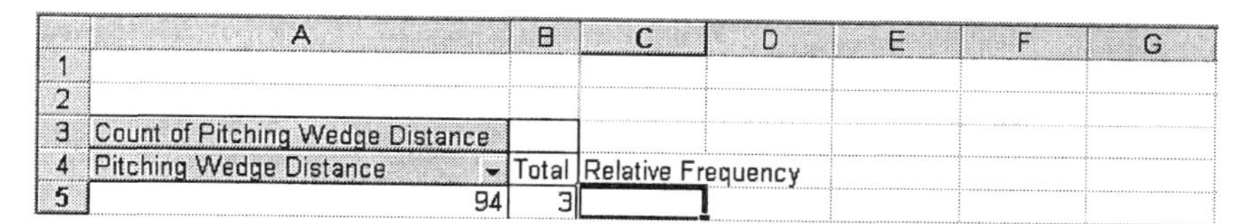

	A	B	C	D	E	F	G
1							
2							
3	Count of Pitching Wedge Distance						
4	Pitching Wedge Distance	Total	Relative Frequency				
5	94	3					

10. Relative frequency is equal to frequency divided by the sum of the frequencies. In cell C5, enter the formula **=B5/$B$19** as shown below. The dollar signs are necessary for the B19 address so that the sum has an absolute reference that will not change when the formula is copied. Press [**Enter**].

	A	B	C	D	E	F	G
1							
2							
3	Count of Pitching Wedge Distance						
4	Pitching Wedge Distance	Total	Relative Frequency				
5	94	3	=B5/$B$19				

11. Copy the contents of cell C5 to cells C6 through C18

	A	B	C	D
1				
2				
3	Count of Pitching Wedge Distance			
4	Pitching Wedge Distance	Total	Relative Frequency	
5	94	3	0.04	
6	95	4	0.053333	
7	96	3	0.04	
8	97	4	0.053333	
9	98	7	0.093333	
10	99	11	0.146667	
11	100	12	0.16	
12	101	12	0.16	
13	102	5	0.066667	
14	103	6	0.08	
15	104	5	0.066667	
16	105	1	0.013333	
17	107	1	0.013333	
18	108	1	0.013333	
19	Grand Total	75		

12. You will now construct a relative frequency line graph. Click and drag over the range **C4:C18** so that these cells are highlighted.

13. At the top of the screen, click **Insert** and select **Chart**.

14. Under Chart type, select **Column**. Under Chart sub-type, select the leftmost diagram in the first row. Click **Next>**.

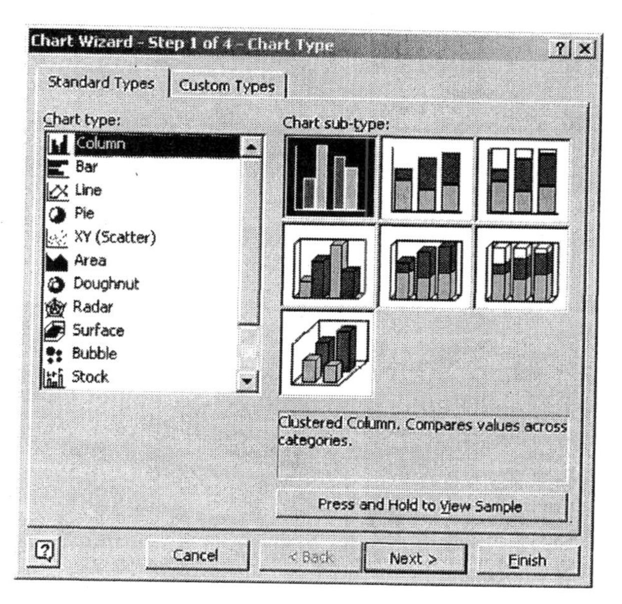

15. You should see =Sheet1!$C$4:$C$18 in the Data range window. Click the **Series** tab at the top of the Chart Source Data dialog box. At the bottom of the dialog box, enter =**Sheet1!$A$5:$A$18** in the Category (X) axis labels window so that the values, 94, 95, 96, etc. appear below the X-axis. Click **Next>**.

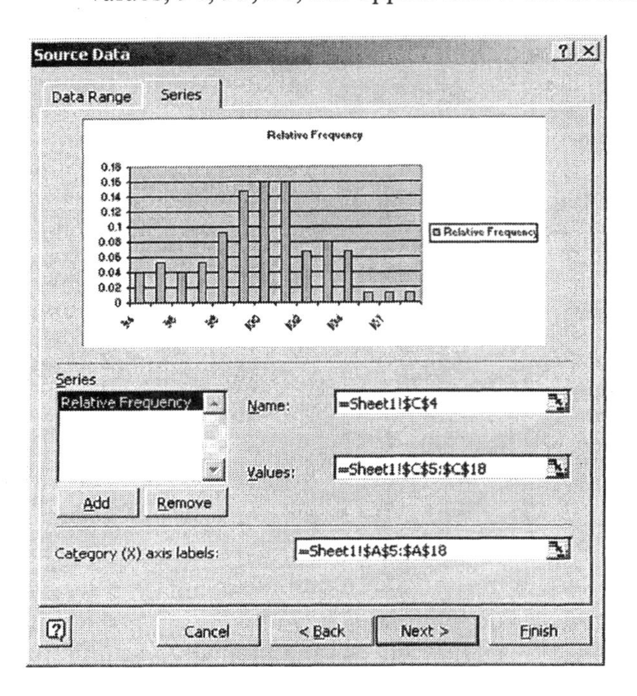

16. Enter a chart title and labels for the X and Y axes. Chart title: **Relative Frequencies of Pitching Wedge Distances**. Category (X) axis: **Pitching Wedge Distance**. Value (Y) axis: **Relative Frequency**. Click **Next>**.

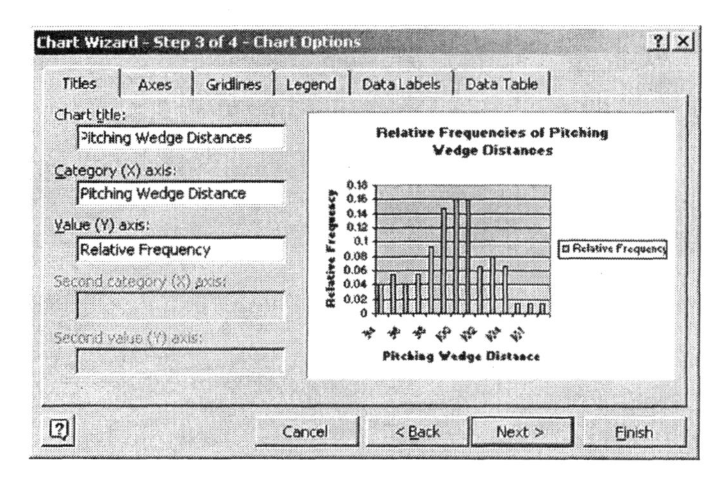

17. In the Chart Location dialog box, select **As new sheet**. Click **Finish**.

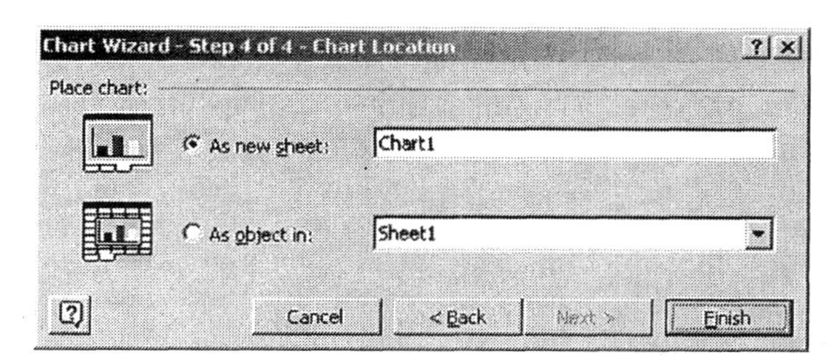

18. Remove the space between the vertical bars. To do this, **right click** directly on one of the vertical bars. Then select **Format Data Series** from the shortcut menu that appears.

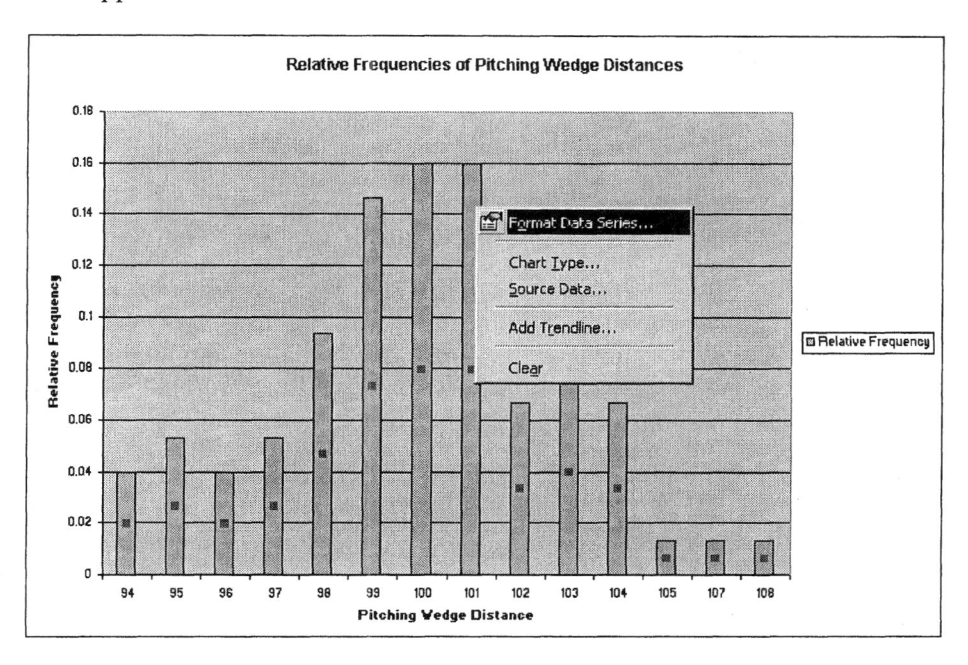

19. Click the **Options** tab at the top of the Format Data Series dialog box. Change the gap width to **0**. Click **OK**. Your completed histogram should look similar to the one shown below.

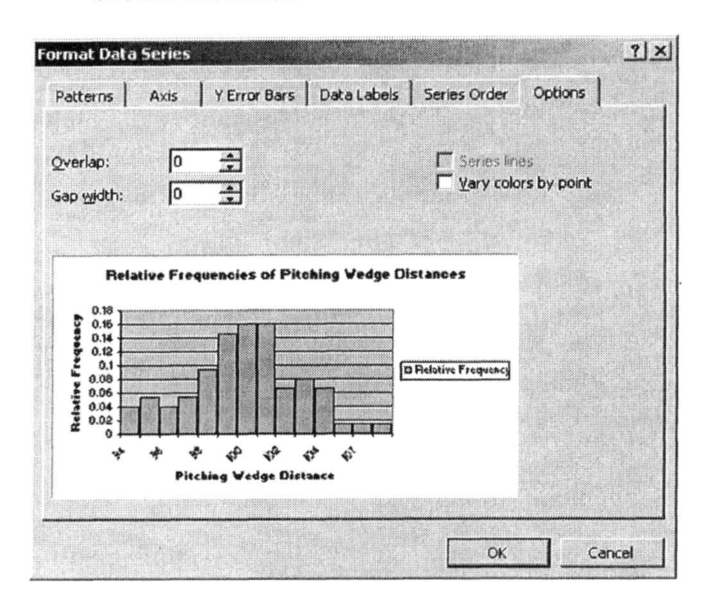

◄

Section 7.2 The Standard Normal Distribution

► Exercise 1 (pg. 403)

Finding the Area Under the Standard Normal Curve

1. Open a new Excel worksheet. Enter the information shown below. You will be using the NORMSDIST function to find the area to the left of specified z values.

	A	B	C	D	E	F	G
1	Z	Area to Left					
2	-2.45						
3	-0.43						
4	1.35						
5	3.49						

2. Click in cell **B2** where you will place the area to the left of z = -2.45. At the top of the screen, click **Insert** and select **Function.**

3. Under Function category, select **Statistical.** Under Function name, select **NORMSDIST**. Note that the NORMSDIST function returns the standard normal cumulative distribution. Click **OK**.

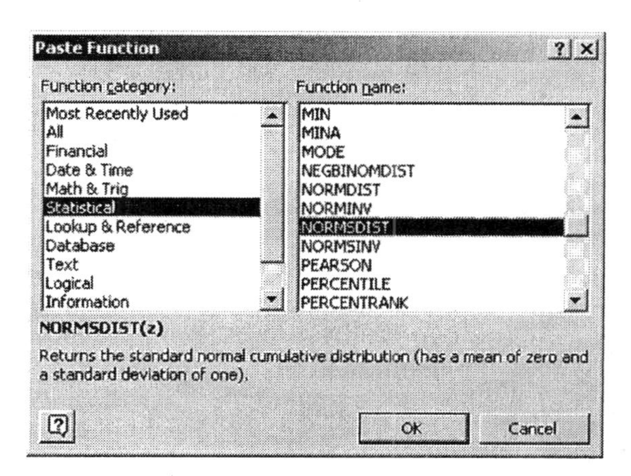

4. Complete the NORMSDIST dialog box as shown below. You are entering the cell address of z = -2.45 rather than the numerical value so that you will be able to copy the function. Click **OK**.

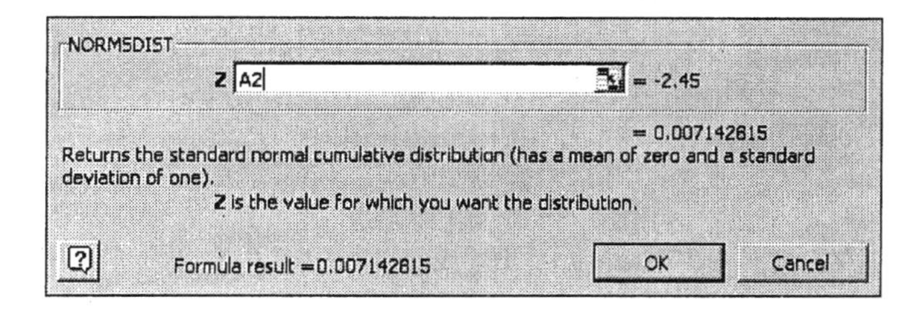

5. Copy the contents of cell B2 to cells B3 through B5. The areas are displayed below.

	A	B	C	D	E	F	G
1	Z	Area to Left					
2	-2.45	0.007143					
3	-0.43	0.333598					
4	1.35	0.911492					
5	3.49	0.999758					

◄

► Exercise 11 (pg. 404) Finding the Z-score that Corresponds to a Specified Area

1. Open a new Excel worksheet. You will be using the NORMSINV function to find the Z-score such that the area under the standard normal curve to the left is 0.1.

2. Click in the cell of the worksheet where you would like to place the output. I clicked in cell **A1**.

3. At the top of the screen, click **Insert** and select **Function**.

4. Under Function category, select **Statistical**. Under Function name, select **NORMSINV**. Click **OK**.

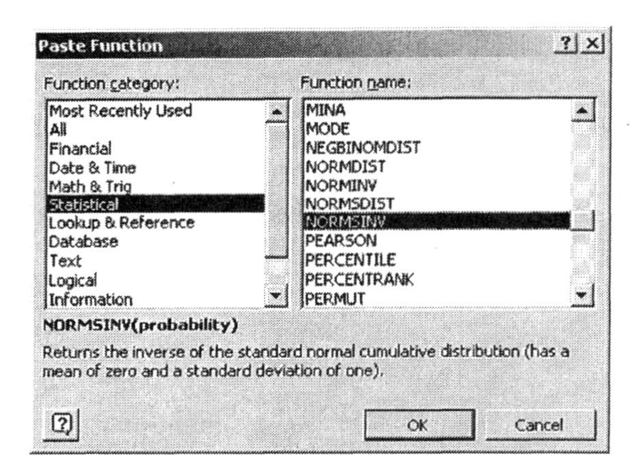

5. Enter the area **.1** in the Probability window. Click **OK**.

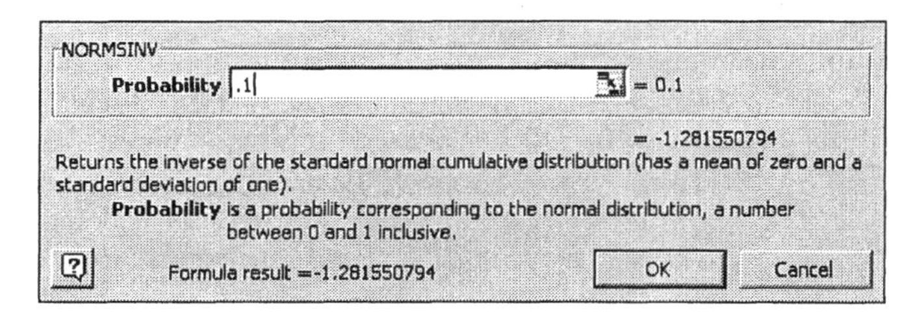

The NORMSINV function returns a Z-score of –1.2816.

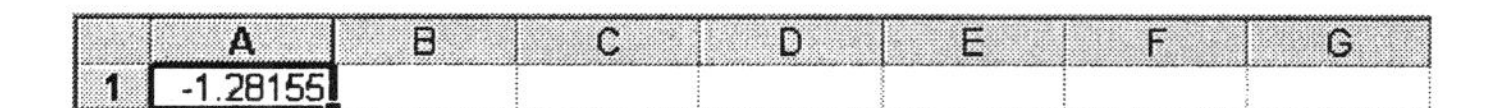

	A	B	C	D	E	F	G
1	-1.28155						

◀

Section 7.3 Applications of the Normal Distribution

► Exercise 11 (pg. 411)

Computing Probabilities of a Normally Distributed Variable

1. Open a new Excel worksheet. Enter the labels as shown below. You will be placing the Z-score in column A and the answers to items (a) through (e) in columns B through F.

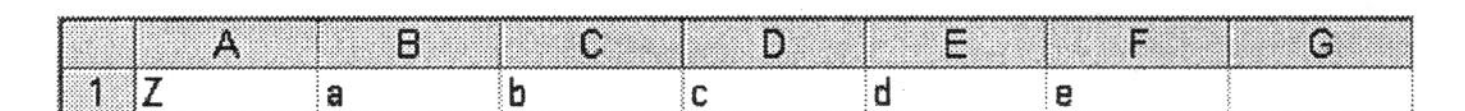

	A	B	C	D	E	F	G
1	Z	a	b	c	d	e	

2. Click in cell **A2** where you will place the Z-score associated with a cholesterol level of 39. At the top of the screen, click **Insert** and select **Function**.

3. Under Function category, select **Statistical**. Under Function name, select **STANDARDIZE**. Click **OK**.

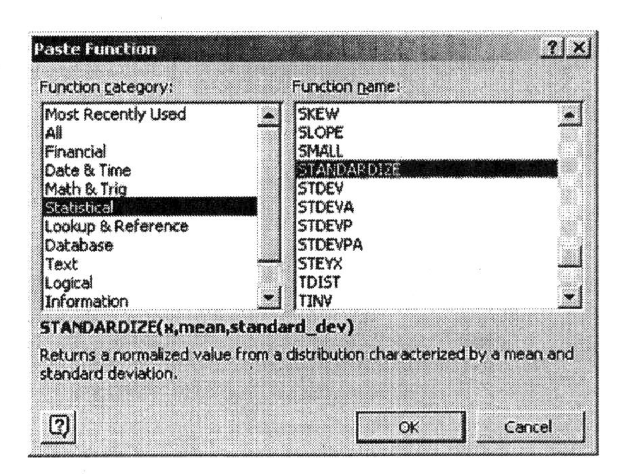

4. Complete the STANDARDIZE dialog box as shown below. Click **OK**.

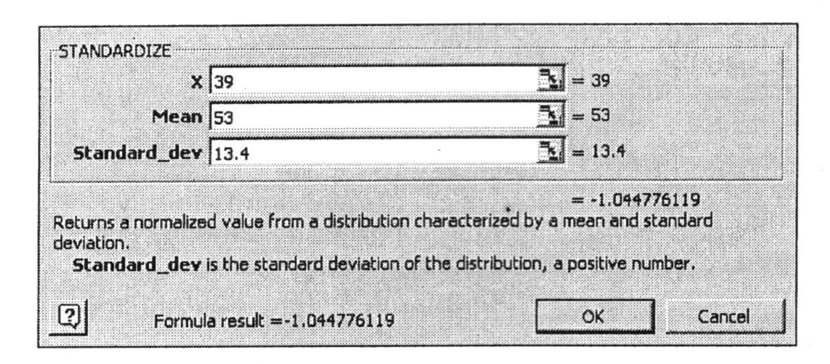

5. The function returns a Z-score of –1.0448. Click in cell **B2** where you will place the answer to item (a). Item (a) can be answered by using the NORMSDIST function which returns the standard normal cumulative distribution.

6. At the top of the screen, click **Insert** and select **Function.**

7. Under Function category, select **Statistical**. Under Function name, select **NORMSDIST**. Click **OK**.

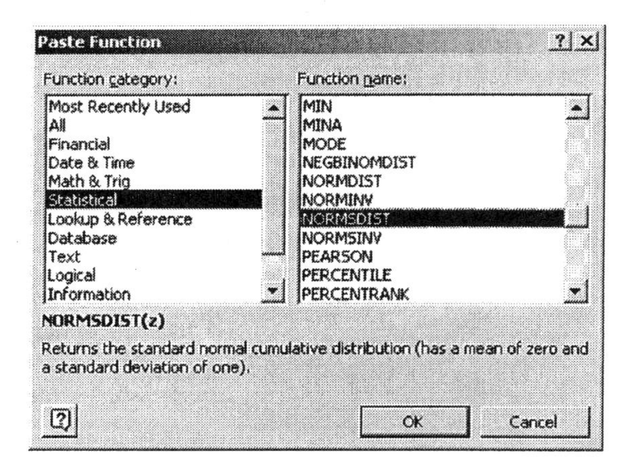

8. Complete the NORMSDIST dialog box as shown below. Click **OK**.

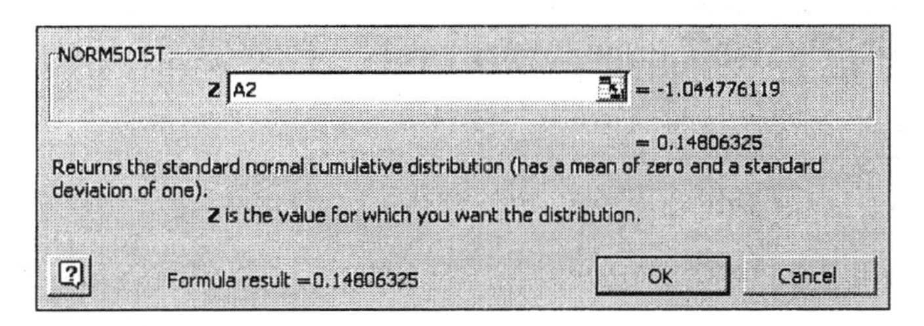

9. The function returns a result of 0.1481. Click in cell **A3** where you will place the Z-score for a cholesterol level of 71. At the top of the screen, click **Insert** and select **Function.**

10. Select **Statistical** and select **STANDARDIZE**. Click OK.

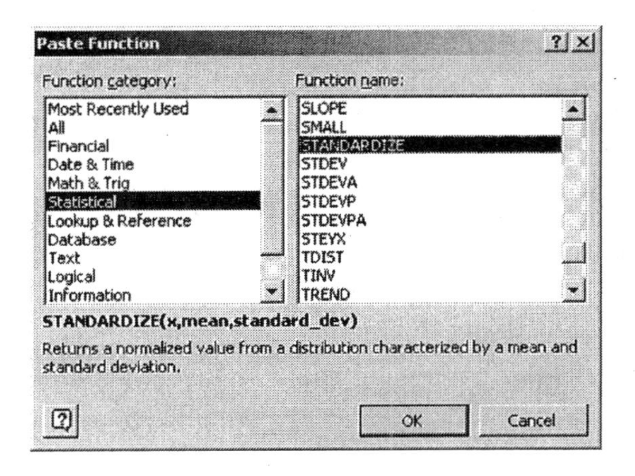

11. Complete the **STANDARDIZE** dialog box as shown below. Click **OK**.

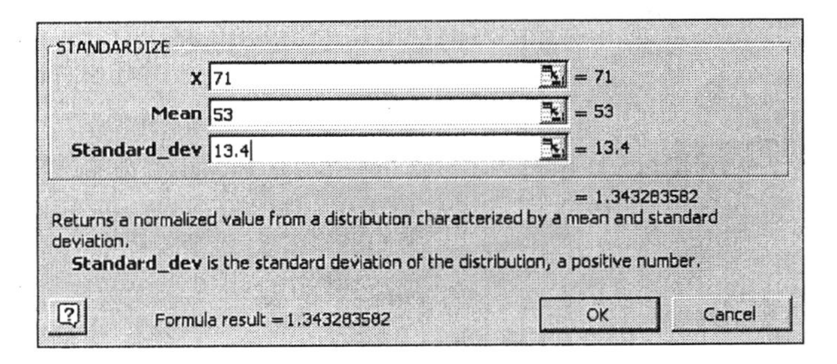

12. The function returns a result of 1.3433. Click in cell **C3** where you will place the answer to item (b). Item (b) is asking for the area to the right of a specified value. Because the NORMSDIST function returns the area to the left of a specific value, you will need to subtract the NORMSDIST result from 1. So, first enter **=1-** in cell C3.

	A	B	C	D	E	F	G
1	Z	a	b	c	d	e	
2	-1.04478	0.148063					
3	1.343284		=1-				

13. Then, click **Insert** and select **Function**.

14. Select **Statistical** and select **NORMSDIST**. Click **OK**.

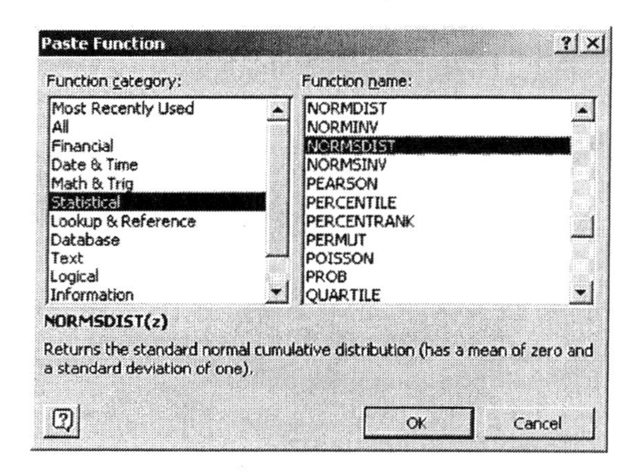

15. Enter the cell address **A3** in the Probability window. (You could also enter 1.343284, but entering the cell address is easier.) Click **OK**.

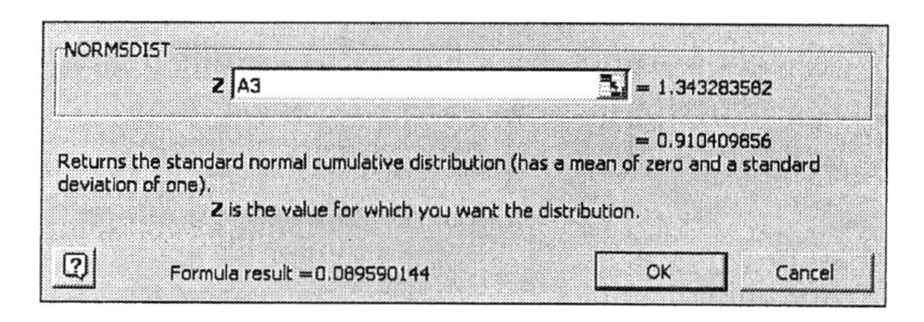

16. The function returns a result of 0.0896. Click in cell **A4** where you will place the Z-score for a cholesterol level of 60. Click **Insert** and select **Function**.

17. Select **Statistical** and select **STANDARDIZE**. Click **OK**.

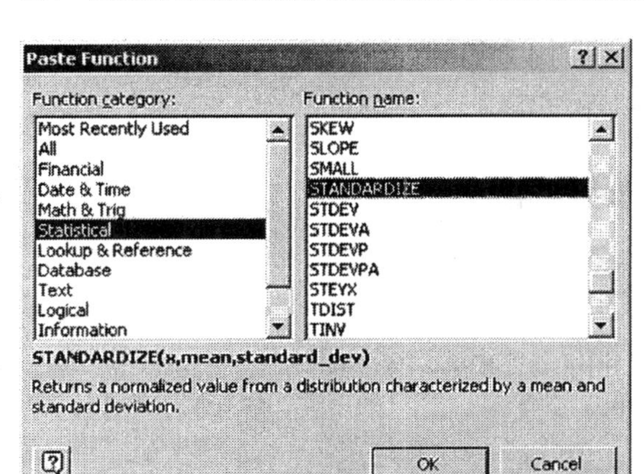

18. Complete the STANDARDIZE dialog box as shown below. Click **OK**.

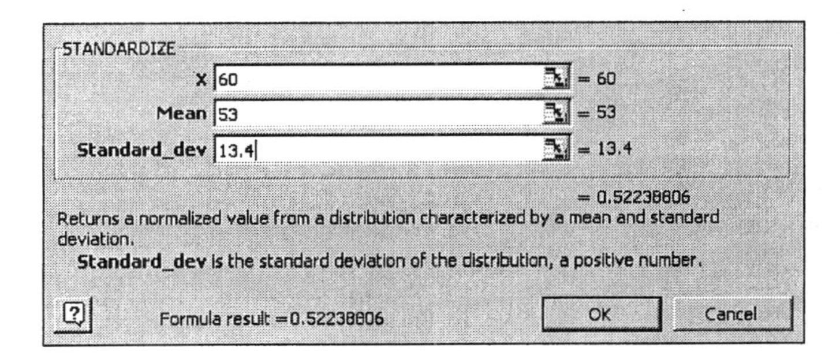

19. The function returns a result of 0.522388. Click in cell **A5** where you will place the Z-score associated with a cholesterol level of 75. Click **Insert** and select **Function**.

20. Select **Statistical** and select **STANDARDIZE**. Click **OK**.

21. Complete the STANDARDIZE dialog box as shown below. Click **OK**.

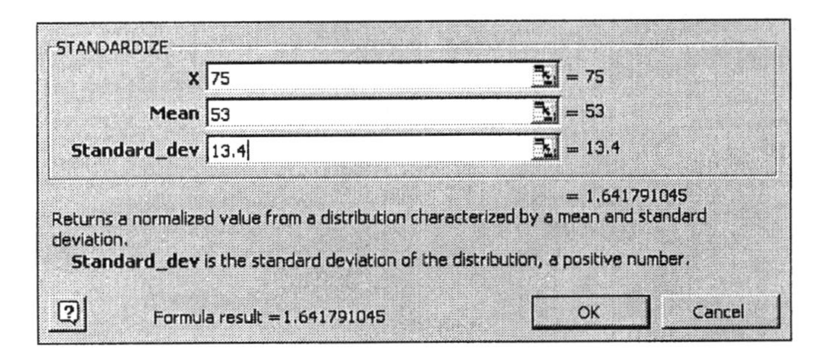

22. The function returns a result of 1.6418. Because NORMSDIST returns the cumulative probability, you can find the area between these two Z-scores by subtracting the cumulative probability of Z = 0.5224 from the cumulative probability of Z = 1.6418. Click in cell **D5** where you will place the answer to item (c). Because this is a formula, you need to begin the cell entry with = as shown below.

	A	B	C	D	E	F	G
1	Z	a	b	c	d	e	
2	-1.04478	0.148063					
3	1.343284		0.08959				
4	0.522388						
5	1.641791			=			

23. Click **Insert** and select **Function**.

24. Select **Statistical** and select **NORMSDIST**. Click **OK**.

25. Complete the NORMSDIST dialog box as shown below. You are requesting the cumulative probability of Z = 1.6418. Click **OK**.

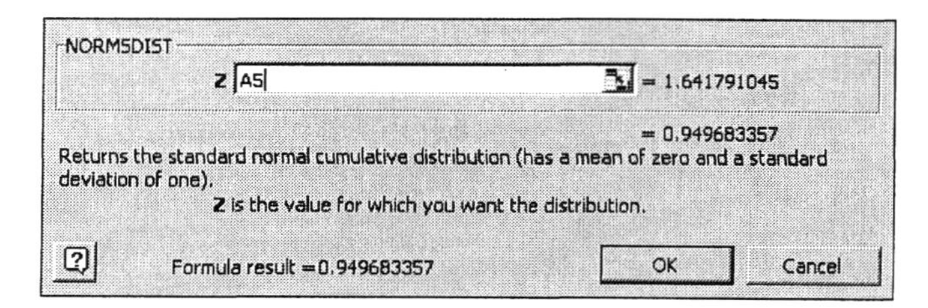

26. To continue the calculation, move the cursor to the formula bar near the top of the screen and click after (A5) so that the I-beam appears there.

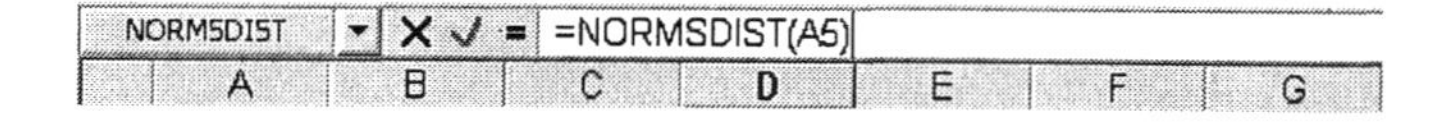

27. Enter a minus sign immediate after (A5). Then click **Insert** and select **Function**. Select **Statistical** and select **NORMSDIST**. Click **OK**.

28. Complete the NORMSDIST dialog box as shown below. Note the complete formula as shown in the formula bar near the top of the screen, =NORMSDIST(A5)-NORMSDIST(A4). Click **OK**.

If you prefer, you could key in this entire formula. You don't have to insert functions from the Function menu.

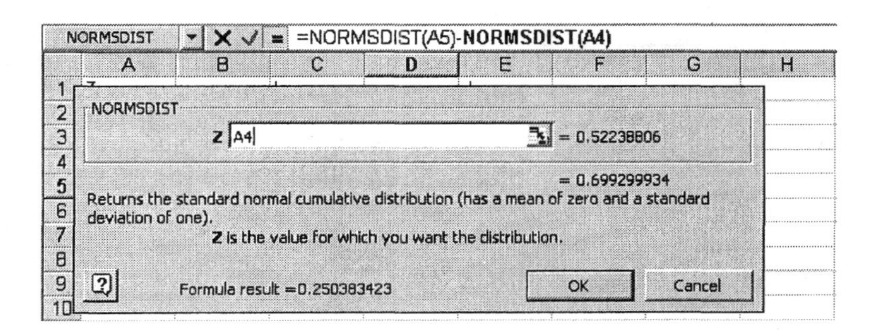

29. The formula returns a result of 0.2504. Click in cell A6 where you will place the Z-score for a cholesterol level of 45. Click **Insert** and select **Function**.

30. Select **Statistical** and select **STANDARDIZE**. Click **OK**.

31. Complete the STANDARDIZE dialog box as shown below. Click **OK**.

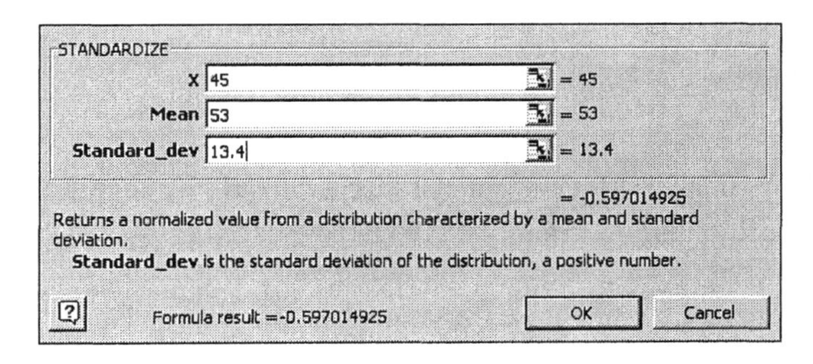

32. The function returns as result of –0.5970. Click in cell **E6** where you will place the answer to item (d). This item can be answered directly by the NORMSDIST function. Click **Insert** and select **Function**.

33. Select **Statistical** and select **NORMSDIST**. Click **OK**.

34. Complete the NORMSDIST dialog box as shown below. Click **OK.**

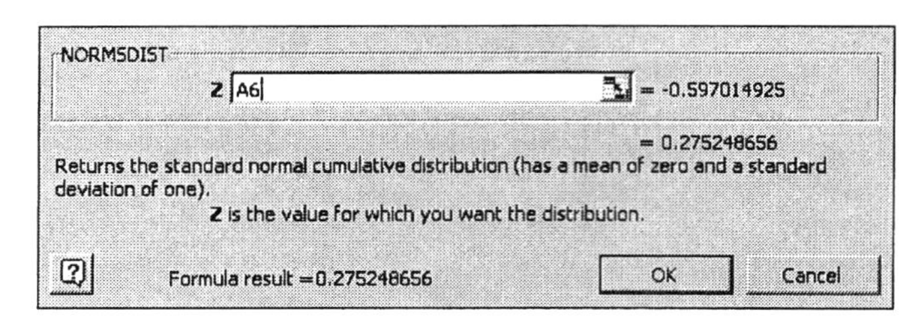

35. The function returns a result of 0.2752. Click in cell **A7** where you will place the Z-score for a cholesterol level of 50. At the top of the screen, click **Insert** and select **Function**.

36. Select **Statistical**. Select **STANDARDIZE**. Click **OK.**

37. Complete the STANDARDIZE dialog box as shown below. Click **OK.**

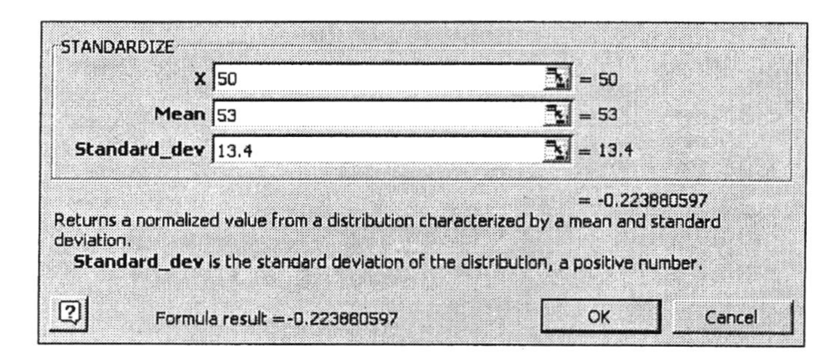

38. The function returns a result of –0.2239. Click in cell **A8** where you will place the Z-score for a cholesterol level of 60. At the top of the screen, click **Insert** and select **Function**.

39. Select **Statistical**. Select **STANDARDIZE**. Click **OK.**

40. Complete the STANDARDIZE dialog box as shown below. Click **OK**.

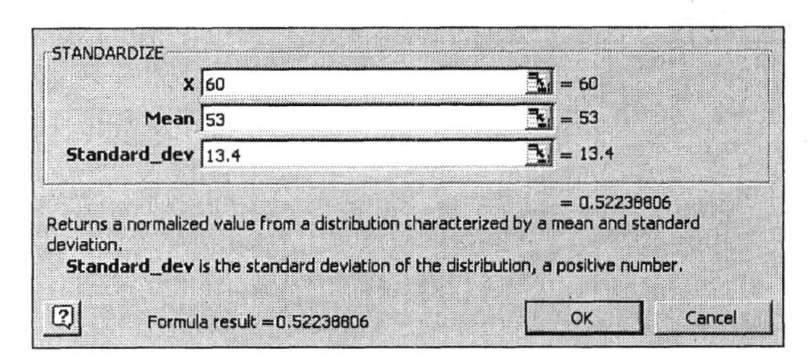

41. The function returns a result of 0.5224. Because NORMSDIST returns the cumulative probability, you can find the area between these two Z-scores by subtracting the cumulative probability of Z = -0.2239 from the cumulative probability of Z = 0.5224. Click in cell **F8** where you will place the answer to item (e). Because this is a formula, you need to begin the cell entry with = as shown below.

	A	B	C	D	E	F	G
1	Z	a	b	c	d	e	
2	-1.04478	0.148063					
3	1.343284		0.08959				
4	0.522388						
5	1.641791			0.250383			
6	-0.59701				0.275249		
7	-0.22388						
8	0.522388					=	

42. Click **Insert** and select **Function**.

43. Select **Statistical** and select **NORMSDIST**. Click **OK**.

44. Complete the NORMSDIST dialog box as shown below. You are requesting the cumulative probability of Z = 0.5224. Click **OK**.

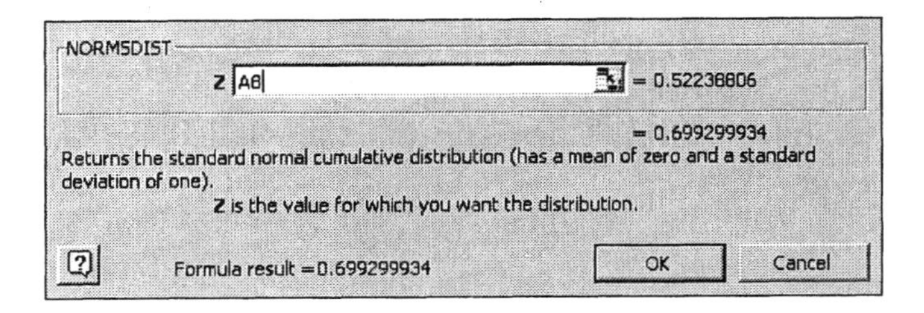

45. To continue the calculation, move the cursor to the formula bar near the top of the screen and click after (A8) so that the I-beam appears there.

NORMSDIST	X ✓ =	=NORMSDIST(A8)				
A	B	C	D	E	F	G

46. Enter a minus sign immediately after (A8). Then click **Insert** and select **Function**. Select **Statistical** and select **NORMSDIST**. Click **OK**.

47. Complete the NORMSDIST dialog box as shown below. You are requesting the cumulative probability of Z = -0.2239. Click **OK**.

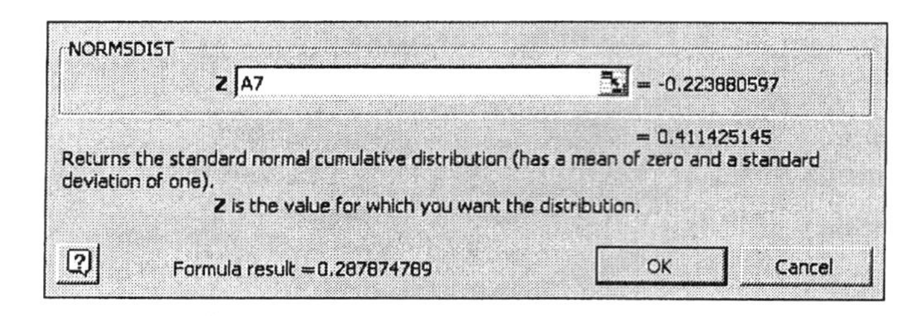

48. The formula returns as result of 0.2879. The completed worksheet is displayed below.

	A	B	C	D	E	F	G
1	Z	a	b	c	d	e	
2	-1.04478	0.148063					
3	1.343284		0.08959				
4	0.522388						
5	1.641791			0.250383			
6	-0.59701				0.275249		
7	-0.22388						
8	0.522388					0.287875	

◄

► Exercise 21 (pg. 412) Finding x That Corresponds to a Specified Percentile

1. Open a new Excel worksheet.

2. Click in cell **A1** where you will place the cholesterol level corresponding to the 25th percentile. You will be using the NORMINV function.

3. At the top of the screen, click **Insert** and select **Function**.

4. Under Function category, select **Statistical**. Under Function name, select **NORMINV**. Click **OK**.

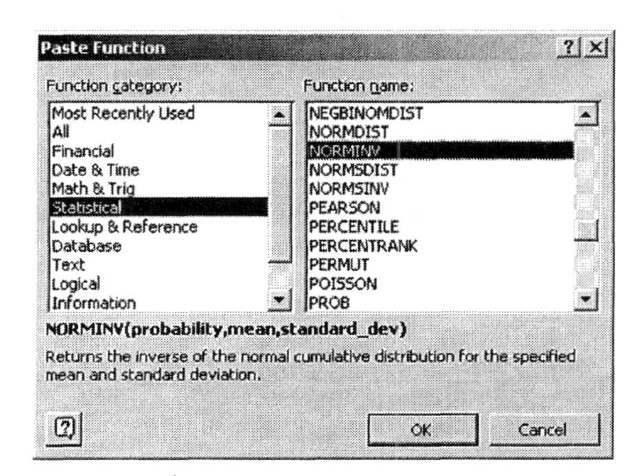

5. Complete the NORMINV dialog box as shown below. Click **OK**.

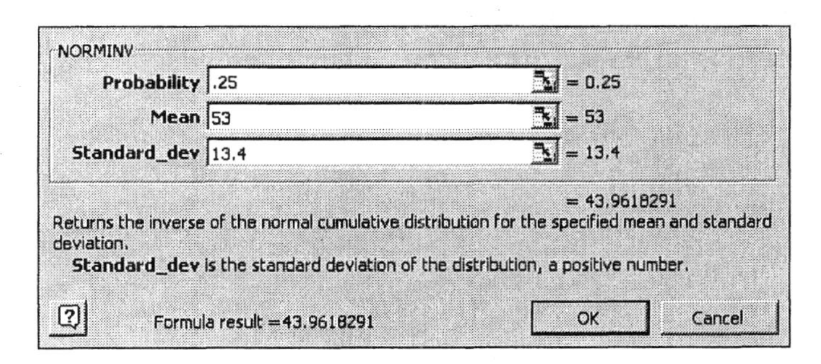

The function returns a result of 43.9618

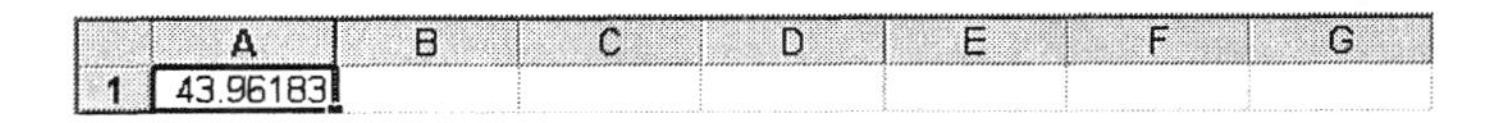

◀

Section 7.4 Assessing Normality

► Example 3 (pg. 419) — Drawing a Normal Probability Plot

1. Open worksheet "7_4_Ex3" in the Chapter 7 folder. The first few lines are shown below.

If the PHStat add-in has not been loaded, you will need to load it before continuing. Follow the instructions in Section GS 8.2

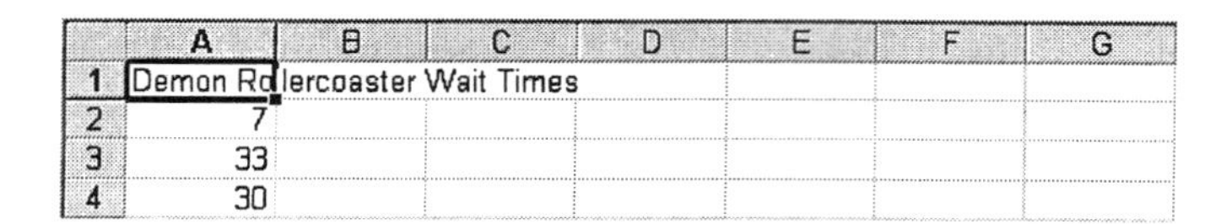

2. At the top of the screen, click **PHStat**. Then select **Probability and Prob. Distributions → Normal Probability Plot**.

3. Complete the Normal Probability Plot dialog box as shown below. Click **OK**.

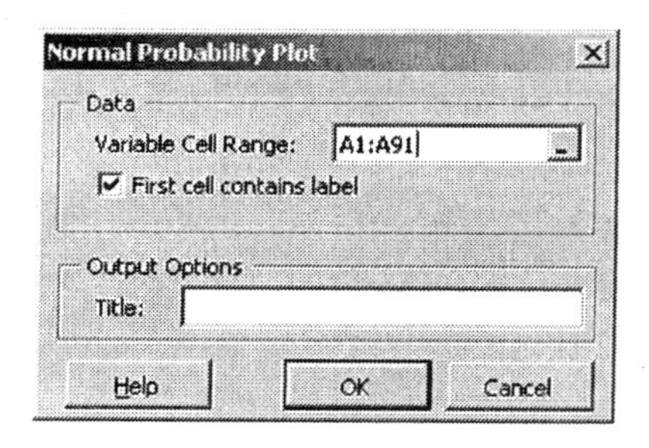

The plot is displayed in a worksheet named NormalPlot. Additional output is provided in a worksheet named Plot.

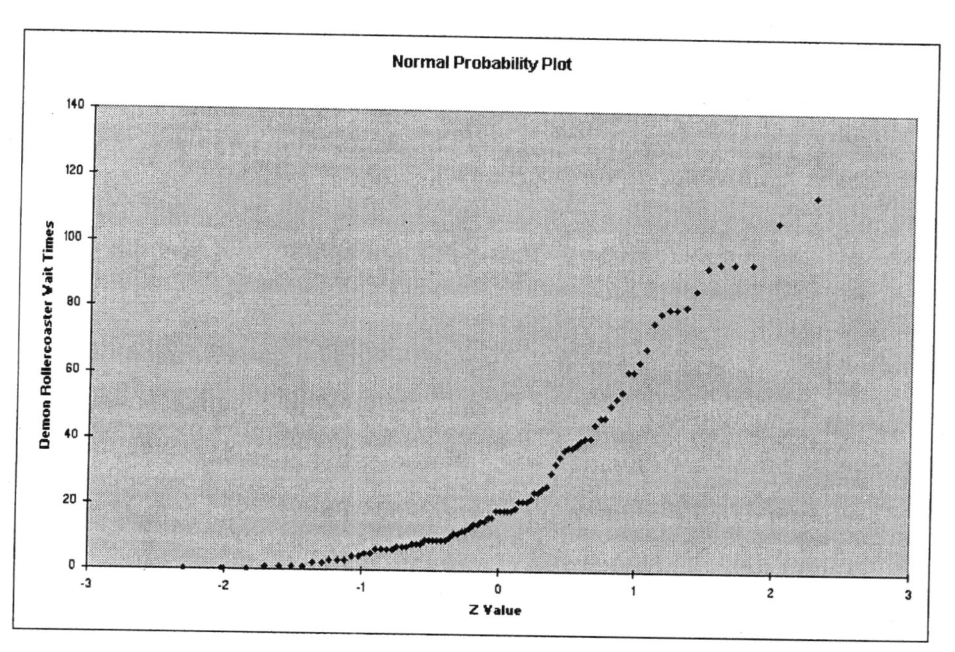

Section 7.5 Sampling Distributions; The Central Limit Theorem

► Exercise 21 (pg. 436) Simulating Scores on the Stanford-Binet IQ Test

1. Open a new Excel worksheet.

If the PHStat add-in has not been loaded, you will need to load it before continuing. Follow the instructions in Section GS 8.2

2. At the top of the screen, click **PHStat**. Select **Sampling → Sampling Distributions Simulation**.

3. Each sample is placed in a separate column of the Excel worksheet. Because the worksheet has only 256 columns, you will ask for 256 samples instead of 500. Complete the Sampling Distributions Simulation dialog box as shown below. Click **OK**.

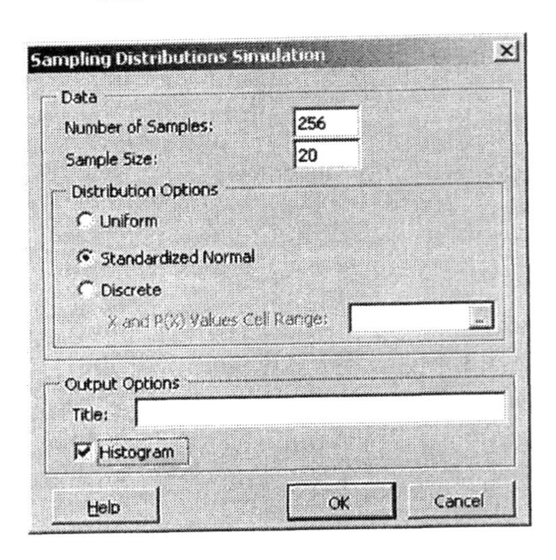

4. This PHStat procedure samples from a standardized normal distribution and generates Z-scores. If you wish to transform the Z-scores to Stanford-Binet IQ test scores, apply the following formula: $X = \mu + (Z)(\sigma)$ where X refers to the IQ score, $\mu = 100$, $\sigma = 16$, and Z refers to the generated Z score.

   The worksheet labeled Histogram contains a histogram of the sample means and a frequency distribution of the sample means.

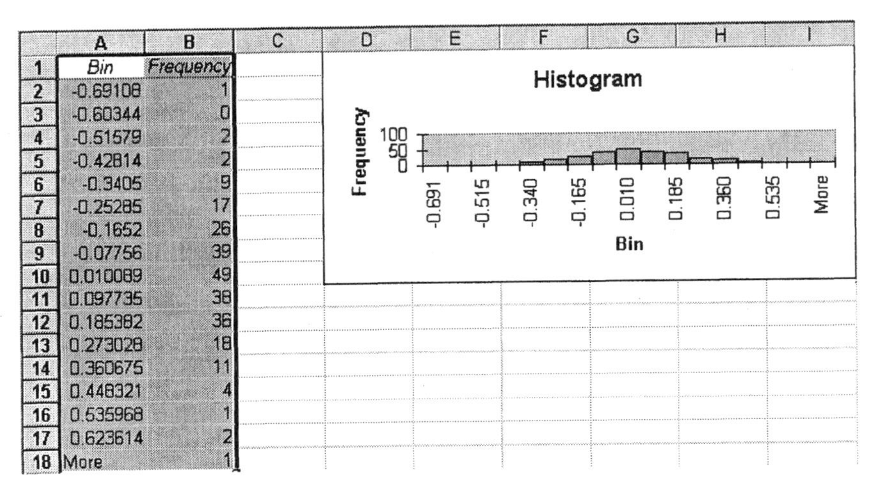

	A	B
1	Bin	Frequency
2	-0.69108	1
3	-0.60344	0
4	-0.51579	2
5	-0.42814	2
6	-0.3405	9
7	-0.25285	17
8	-0.1652	26
9	-0.07756	39
10	0.010089	49
11	0.097735	38
12	0.185382	36
13	0.273028	18
14	0.360675	11
15	0.448321	4
16	0.535968	1
17	0.623614	2
18	More	1

5. To make the chart taller so that it is easier to read, click within the figure near a border so that black square handles appear. Then, click on the center handle on the bottom border of the figure and drag it down a few rows.

6. Click on the **SDS** sheet tab at the bottom of the screen. This worksheet presents the scores in each of the 256 samples of size n = 2, the Sample Means, the Overall Average, and the Standard Error of the Mean. Let's transform the means from Z-values to Stanford-Binet IQ test values. First, enter labels for your output as shown below.

23	Sample Means:						
24	-0.13017	-0.15769	0.135946	0.074014	-0.05523	0.007863	-0.08279
25	Overall Average:						
26	-0.01891						
27	Standard Error of the Mean:						
28	0.206639						
29	Sample Means:						
30							
31	Overall Average:						
32							
33	Standard Error of the Mean:						
34							

7. Click in cell **A30** where you will place the IQ test value corresponding to the Z-value displayed in A24 (-0.1302). Enter the formula **=100+A24*16** as shown below. Press **[Enter]**.

23	Sample Means:						
24	-0.13017	-0.15769	0.135946	0.074014	-0.05523	0.007863	-0.08279
25	Overall Average:						
26	-0.01891						
27	Standard Error of the Mean:						
28	0.206639						
29	Sample Means:						
30	=100+A24*16						

8. Copy the formula in A30 all the way across from cell B30 to the last column in the worksheet, IV30. Row 30 of the worksheet now contains the means expressed as Stanford-Binet IQ values. Next, click in cell **A32** where you will place the overall average of these means. At the top of the screen, click **Insert** and select **Function**.

9. Under Function category, select **Statistical**. Under Function name, select **AVERAGE**. Click **OK**.

10. Complete the AVERAGE dialog box as shown below. Click **OK**.

The range 30:30 indicates all the values in row 30 of the worksheet.

AVERAGE
Number1 30:30 = {97.9172434905195
Number2 = number
= 99.69746665
Returns the average (arithmetic mean) of its arguments, which can be numbers or names, arrays, or references that contain numbers.
Number1: number1,number2,... are 1 to 30 numeric arguments for which you want the average.
Formula result =99.69746665
OK Cancel

11. The function returns a value of 99.6975. Click in cell **A34** where you will place the standard deviation of the means (i.e., the standard error of the mean). At the top of the screen, click **Insert** and select **Function**.

12. Under Function category, select **Statistical**. Under Function name, select **STDEVP**. Click **OK**.

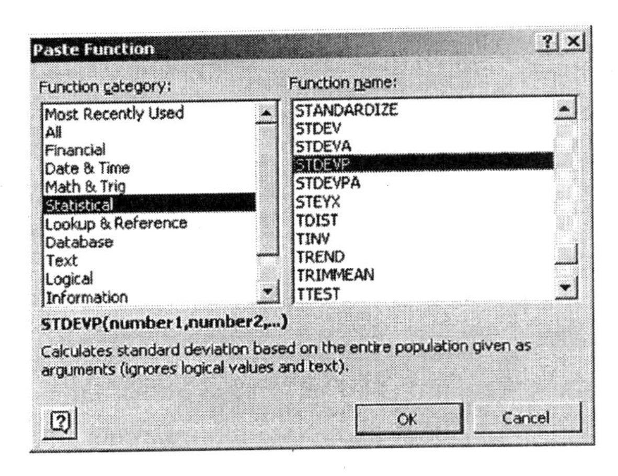

13. Complete the STDEVP dialog box as shown below. Click **OK**.

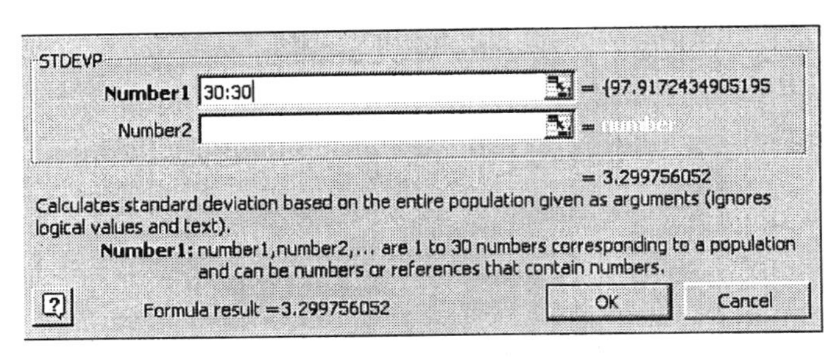

The function returns a value of 3.2998. Because the samples were generated randomly, it is not likely that your result will be exactly the same.

29	Sample Means:						
30	97.91724	97.47691	102.1751	101.1842	99.11637	100.1258	98.6753
31	Overall Average:						
32	99.69747						
33	Standard Error of the Mean:						
34	3.299756						

◀

Confidence Intervals

CHAPTER 8

Section 8.1 Confidence Intervals about a Population Mean, σ Known

► Exercise 15 (pg. 468)

Constructing 95% and 90% Confidence Intervals

1. Open a new Excel worksheet and enter the costs as shown below.

	A	B	C	D	E	F	G
1	Cost of Repairs						
2	225						
3	462						
4	729						
5	753						

If the PHStat add-in has not been loaded, you will need to load it before continuing. Follow the instructions in Section GS 8.2.

2. At the top of the screen, click **PHStat** and select **Confidence Intervals → Estimate for the Mean, sigma known**.

3. First, you will request the 95% confidence level. Complete the Estimate for the Mean dialog box as shown below. Click **OK**.

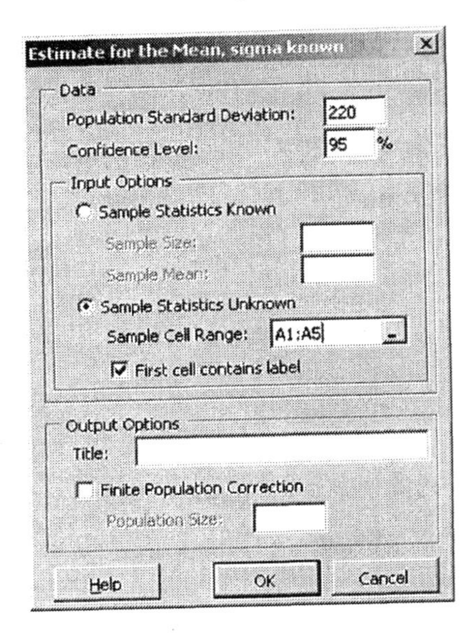

4. The output is placed in a worksheet named Confidence. At the bottom of the screen, click on the **Sheet1** tab to return to the data.

	A	B	C	D	E	F	G
1	**Confidence Interval Estimate for the Mean**						
2							
3	**Data**						
4	**Population Standard Deviation**	220					
5	**Sample Mean**	542.25					
6	**Sample Size**	4					
7	**Confidence Level**	95%					
8							
9	Intermediate Calculations						
10	Standard Error of the Mean	110					
11	Z Value	-1.95996108					
12	Interval Half Width	215.5957191					
13							
14	**Confidence Interval**						
15	**Interval Lower Limit**	326.6542809					
16	**Interval Upper Limit**	757.8457191					

5. Repeat these steps to obtain the 90% confidence level. At the top of the screen, click **PHStat**. Select **Confidence Intervals → Estimate for the mean, sigma known**.

6. Complete the Estimate for the Mean dialog box as shown below. Click **OK**.

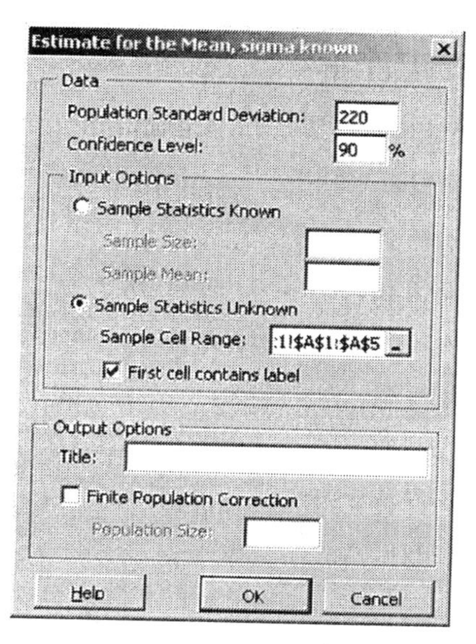

The output is placed in a worksheet named Confidence 2.

	A	B	C	D	E	F	G
1	**Confidence Interval Estimate for the Mean**						
2							
3	**Data**						
4	**Population Standard Deviation**	220					
5	**Sample Mean**	542.25					
6	**Sample Size**	4					
7	**Confidence Level**	90%					
8							
9	Intermediate Calculations						
10	Standard Error of the Mean	110					
11	Z Value	-1.644853					
12	Interval Half Width	180.9338301					
13							
14	**Confidence Interval**						
15	**Interval Lower Limit**	361.3161699					
16	**Interval Upper Limit**	723.1838301					

Section 8.2 Confidence Intervals about a Population Mean, σ Unknown

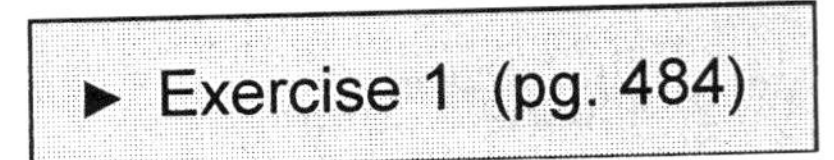

Finding t-Values

1. Open a new Excel worksheet and enter labels for items a, b, c, and d as shown below.

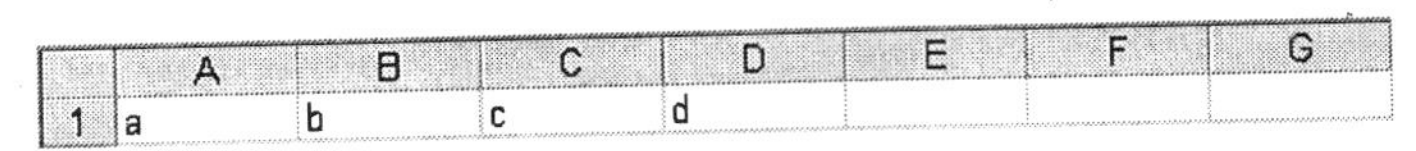

2. Click in cell **A2** where you will place the answer to item a. At the top of the screen, click **Insert** and select **Function**.

3. Under Function category, select **Statistical**. Under Function name, select **TINV**.

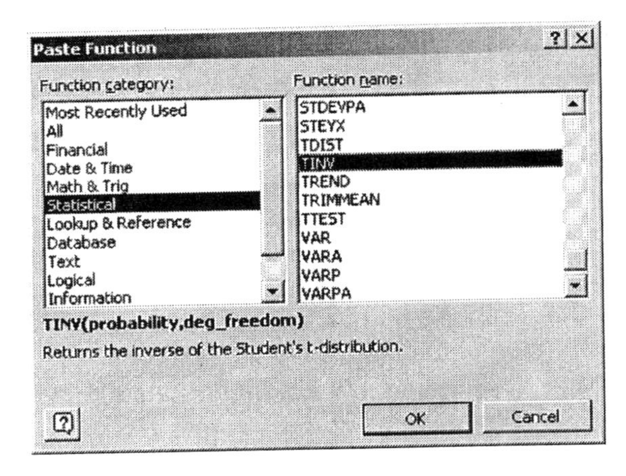

4. Complete the TINV dialog box as shown below. The probability refers to a two-tailed probability. So, to get obtain the t-value for area of 0.10 in the right tail alone, request probability of **.20**, i.e., 0.10 in each tail. Click **OK**.

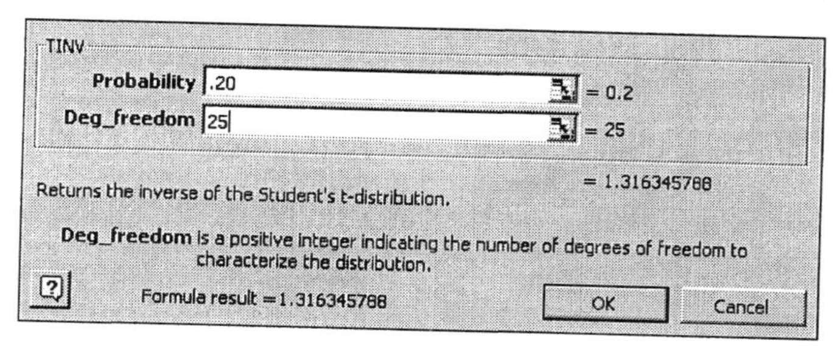

5. The function returns a value of 1.3163. Click in cell **B2** where you will place the answer to item b. At the top of the screen, click **Insert** and select **Function**.

6. Select **Statistical** and select **TINV**. Click **OK**.

7. Complete the TINV dialog box as shown below. The probability refers to a two-tailed probability. So, to get obtain the t-value for area of 0.05 in the right tail alone, request probability of **.10**, i.e., 0.05 in each tail. Click **OK**.

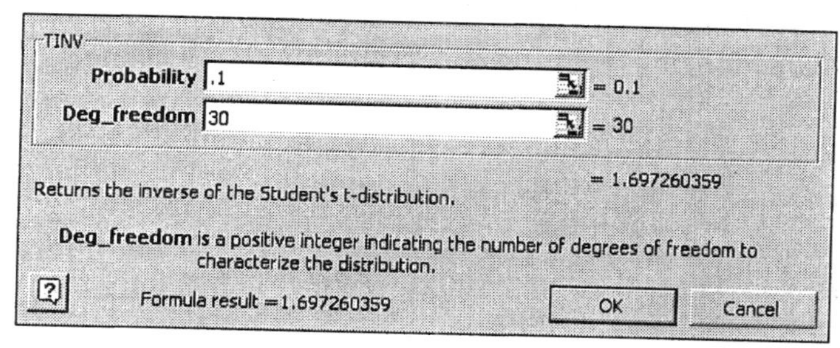

8. The function returns a value of 1.6973. Click in cell **C2** where you will place the answer to item C.

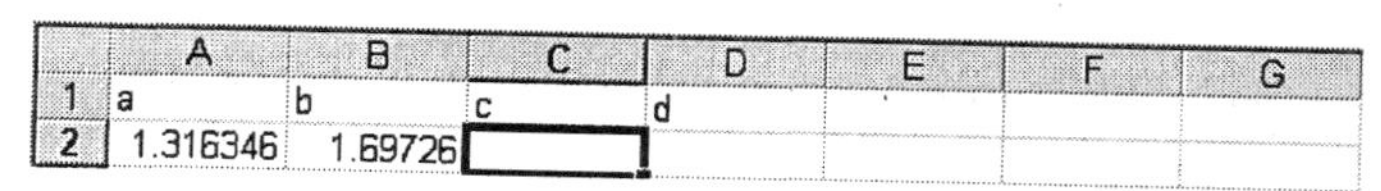

	A	B	C	D	E	F	G
1	a	b	c	d			
2	1.316346	1.69726					

9. At the top of the screen, click **Insert** and select **Function**.

10. Under Function category, select **Statistical**. Under Function name, select **TINV**. Click **OK**.

11. Complete the TINV dialog box as shown below. The probability refers to a two-tailed probability. So, to get obtain the t-value for area of 0.01 in the left tail, request probability of 0.02, i.e., 0.01 in each tail.

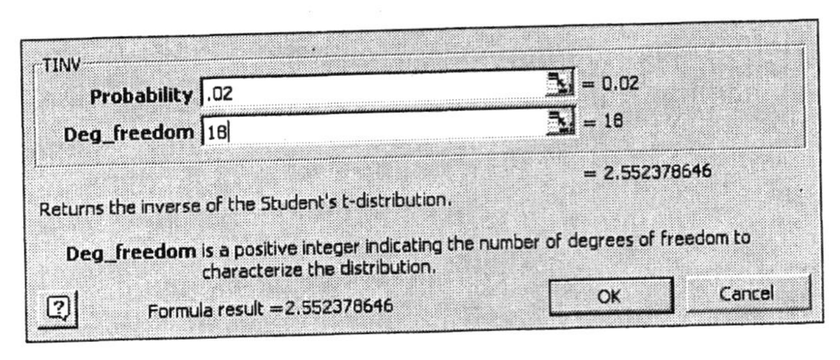

12. The function returns a value of 2.5524. Because the function returns the absolute value, and the t-value you need is in the left tail, the correct result is negative, -2.5524. Click in cell **D2** where you will place the answer to item d.

13. At the top of the screen, click **Insert** and select **Function**.

14. Under Function category, select **Statistical**. Under Function name, select **TINV**. Click **OK**.

15. Complete the TINV dialog box as shown below. To obtain the t-value corresponding to 90% confidence, you input a two-tailed probability of **.10** (i.e., 0.05 in each tail). Click **OK**.

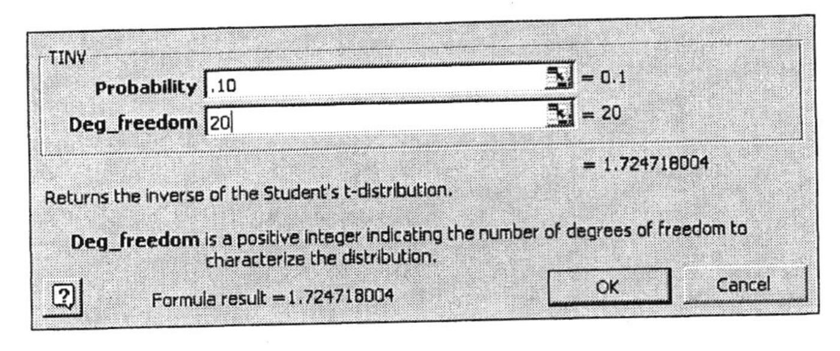

The function returns a value of 1.7247. The completed worksheet is displayed below.

	A	B	C	D	E	F	G
1	a	b	c	d			
2	1.316346	1.69726	2.552379	1.724718			

◀

► Exercise 3 (pg. 484) — Constructing 96% and 90% Confidence Intervals

1. Open a new Excel worksheet.

If the PHStat add-in has not been loaded, you will need to load it before continuing. Follow the instructions in Section GS 8.2.

---

2. At the top of the screen, click **PHStat**. Select **Confidence Intervals → Estimate for the Mean, sigma unknown**.

3. To construct the 96% confidence interval for n = 25, complete the Estimate for the Mean dialog box as shown below. Click **OK**.

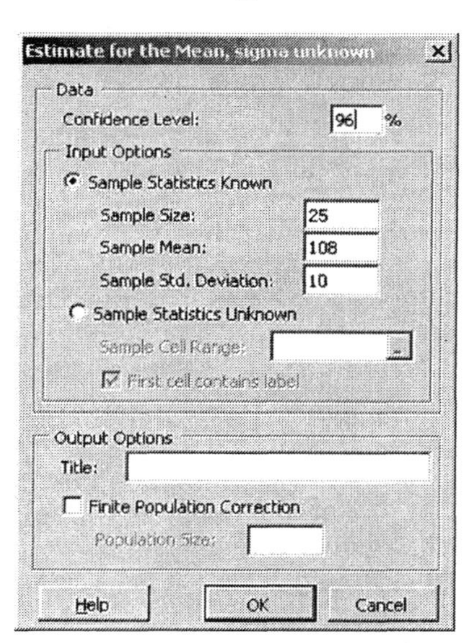

4. The output is displayed in a worksheet named Confidence. To construct the 96% confidence interval for n = 10, begin by clicking **PHStat** at the top of the screen. Select **Confidence Intervals → Estimate for the Mean, sigma unknown**.

	A	B	C	D	E	F	G
1	**Confidence Interval Estimate for the Mean**						
2							
3	**Data**						
4	**Sample Standard Deviation**	10					
5	**Sample Mean**	108					
6	**Sample Size**	25					
7	**Confidence Level**	96%					
8							
9	Intermediate Calculations						
10	Standard Error of the Mean	2					
11	Degrees of Freedom	24					
12	*t* Value	2.17154593					
13	Interval Half Width	4.343091859					
14							
15	**Confidence Interval**						
16	**Interval Lower Limit**	103.66					
17	**Interval Upper Limit**	112.34					

5. Complete the Estimate for the Mean dialog box as shown below. Click **OK**.

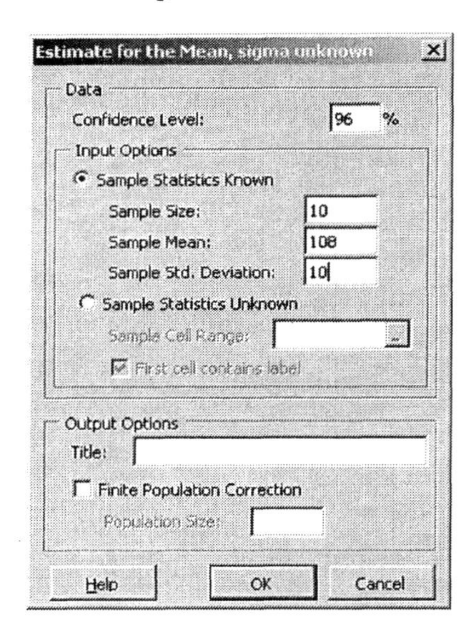

6. The output is displayed in a worksheet named Confidence2. To construct the 90% confidence interval with n = 25, begin by clicking **PHStat** at the top of the screen. Select **Confidence Intervals → Estimate for the Mean, sigma unknown**.

7. Complete the Estimate for the Mean dialog box as shown below. Click **OK**.

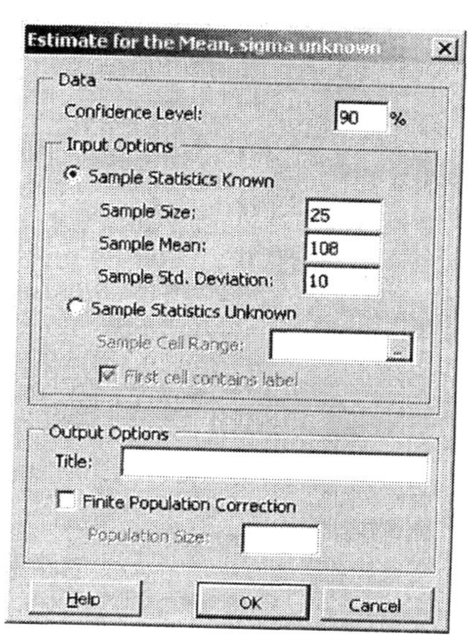

The output is displayed in a worksheet named Confidence3.

	A	B	C	D	E	F	G
1	**Confidence Interval Estimate for the Mean**						
2							
3	**Data**						
4	**Sample Standard Deviation**	**10**					
5	**Sample Mean**	**108**					
6	**Sample Size**	**25**					
7	**Confidence Level**	**90%**					
8							
9	Intermediate Calculations						
10	Standard Error of the Mean	2					
11	Degrees of Freedom	24					
12	*t* Value	1.710882316					
13	Interval Half Width	3.421764632					
14							
15	**Confidence Interval**						
16	**Interval Lower Limit**	**104.58**					
17	**Interval Upper Limit**	**111.42**					

◀

▶ Exercise 25 (pg. 489) Constructing a Boxplot and Confidence Intervals

1. Open a new Excel worksheet and enter the price data as shown below.

If the PHStat add-in has not been loaded, you will need to load it before continuing. Follow the instructions in Section GS 8.2.

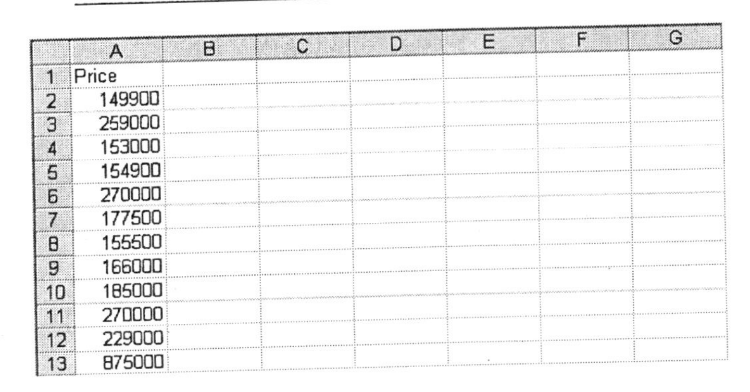

	A	B	C	D	E	F	G
1	Price						
2	149900						
3	259000						
4	153000						
5	154900						
6	270000						
7	177500						
8	155500						
9	166000						
10	185000						
11	270000						
12	229000						
13	875000						

2. At the top of the screen, click **PHStat** and select **Descriptive Statistics → Box-and-Whisker Plot**.

3. Complete the Box-and-Whisker Plot dialog box as shown below. Click **OK**.

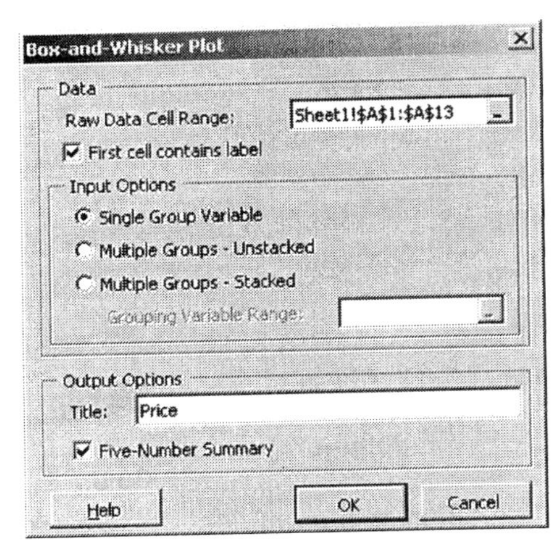

4. The box-and-whisker plot is displayed in a worksheet named BoxWhiskerPlot. Return to the worksheet containing the data. To do this, click on the **Sheet1** tab near the bottom of the screen.

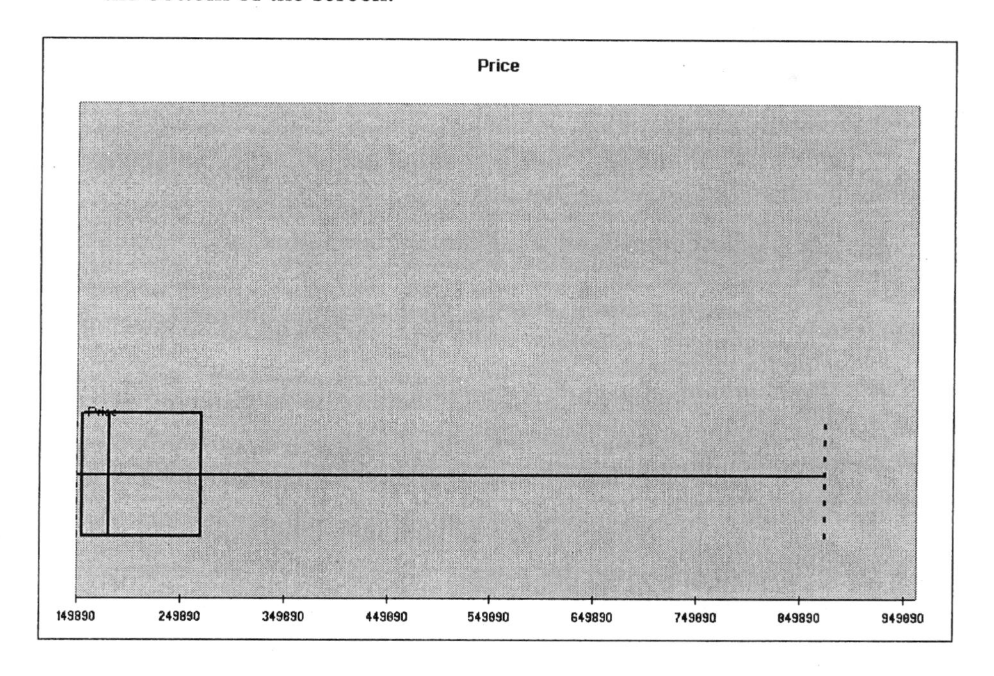

5. The outlier is 875,000. Next you will construct a 99% confidence interval that includes this outlier. At the top of the screen, click **PHStat**. Select **Confidence Intervals → Estimate for the Mean, sigma unknown**.

6. Complete the Estimate for the Mean dialog box as shown below. Click **OK**.

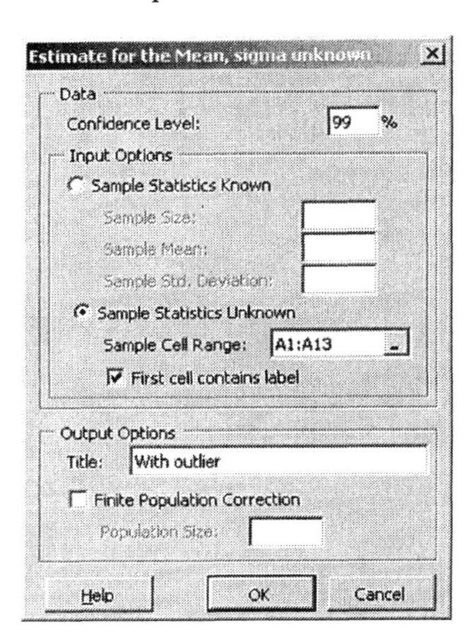

7. The output is placed in a worksheet named Confidence. You now will construct a 99% confidence interval that does not include the outlier. Return to the sheet that contains the data by clicking on the **Sheet1** tab at the bottom of the screen. At the top of the screen, click PHStat. Select **Confidence Intervals → Estimate for the Mean, sigma unknown**.

	A	B	C	D	E	F	G
1	**With outlier**						
2							
3	**Data**						
4	**Sample Standard Deviation**	**201295.8625**					
5	**Sample Mean**	**253733.3333**					
6	**Sample Size**	**12**					
7	**Confidence Level**	**99%**					
8							
9	Intermediate Calculations						
10	Standard Error of the Mean	58109.11019					
11	Degrees of Freedom	11					
12	*t* Value	3.105815267					
13	Interval Half Width	180476.1616					
14							
15	**Confidence Interval**						
16	**Interval Lower Limit**	**73257.17**					
17	**Interval Upper Limit**	**434209.49**					

8. Complete the Estimate for the Mean dialog box as shown below. Note that the range excludes 875,000. Click **OK**.

The complete Sample Cell Range is Sheet1!:$A$1:$A$12. You could also type A1:A12.

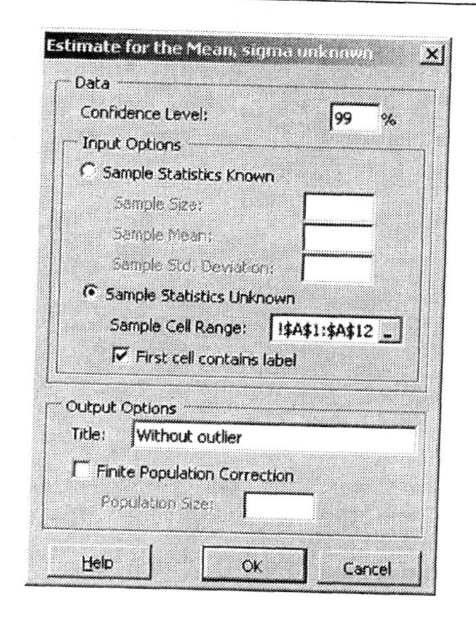

The output is placed in a worksheet named Confidence2.

	A	B	C	D	E	F	G
1	**Without outlier**						
2							
3	**Data**						
4	**Sample Standard Deviation**	**49658.44064**					
5	**Sample Mean**	**197254.5455**					
6	**Sample Size**	**11**					
7	**Confidence Level**	**99%**					
8							
9	Intermediate Calculations						
10	Standard Error of the Mean	14972.58321					
11	Degrees of Freedom	10					
12	*t* Value	3.169261618					
13	Interval Half Width	47452.03328					
14							
15	**Confidence Interval**						
16	**Interval Lower Limit**	**149802.51**					
17	**Interval Upper Limit**	**244706.58**					

◀

Section 8.3 Confidence Intervals about a Population Proportion

► Exercise 7 (pg. 495) — Estimating a Population Proportion and Constructing a 90% Confidence Interval

If the PHStat add-in has not been loaded, you will need to load it before continuing. Follow the instructions in Section GS 8.2.

1. At the top of the screen, click **PHStat**. Select **Confidence Intervals → Estimate for the Proportion**.

2. Complete the Estimate for the Proportion dialog box as shown below. Click **OK**.

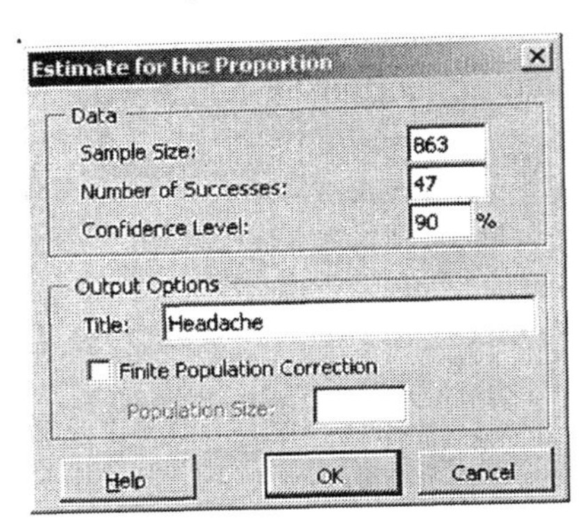

The output is placed in a worksheet named Confidence.

	A	B	C	D	E	F	G
1	Headache						
2							
3	Data						
4	Sample Size	863					
5	Number of Successes	47					
6	Confidence Level	90%					
7							
8	Intermediate Calculations						
9	Sample Proportion	0.054461182					
10	Z Value	-1.644853					
11	Standard Error of the Proportion	0.007724632					
12	Interval Half Width	0.012705885					
13							
14	Confidence Interval						
15	Interval Lower Limit	0.041755297					
16	Interval Upper Limit	0.067167067					

◀

Section 8.4 Confidence Intervals about a Population Standard Deviation

▶ Exercise 9 (pg. 504) Constructing a 99% Confidence Interval about a Population Standard Deviation

If the PHStat add-in has not been loaded, you will need to load it before continuing. Follow the instructions in Section GS 8.2.

1. Open worksheet "8_4_9" in the Chapter 8 folder. The first few lines are shown below.

	A	B	C	D	E	F	G
1	Tensile Strength						
2	203.41						
3	209.58						
4	213.35						

2. The confidence interval procedure will ask you for the sample size and the sample standard deviation. Use Excel's Data Analysis tool to obtain these values. At the top of the screen, click **Tools** and select **Data Analysis.**

3. Select **Descriptive Statistics** and click **OK**.

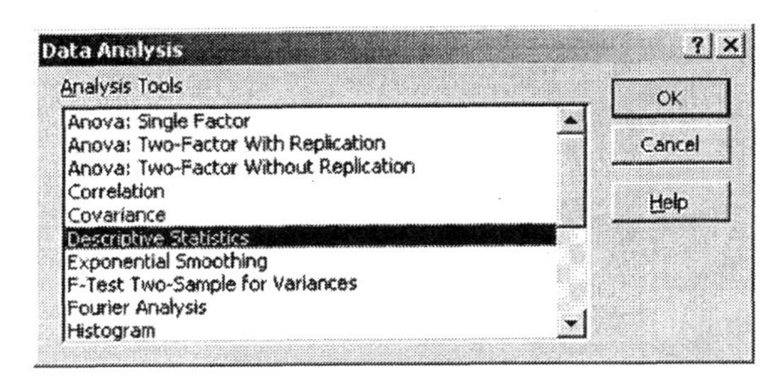

4. Complete the Descriptive Statistics dialog box as shown below. Click **OK**.

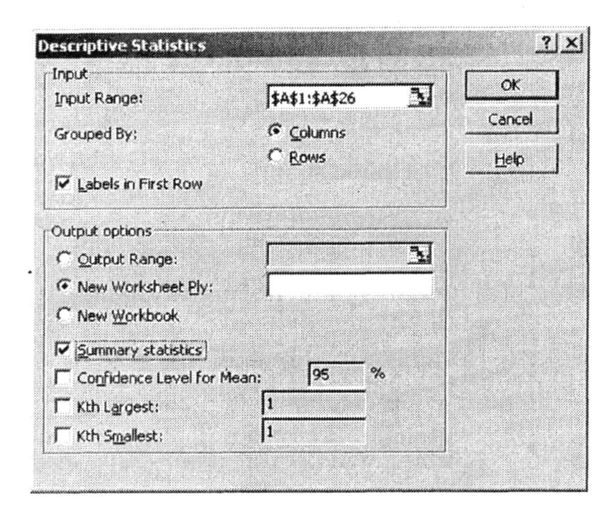

5. Make the column A wider so that you can read all the labels. The standard deviation is 18.3204 and the sample size (Count) is 25.

	A	B	C	D	E	F	G
1	*Tensile Strength*						
2							
3	Mean	203.9336					
4	Standard Error	3.664086					
5	Median	206.51					
6	Mode	#N/A					
7	Standard Deviation	18.32043					
8	Sample Variance	335.6381					
9	Kurtosis	0.468697					
10	Skewness	-0.31621					
11	Range	82.32					
12	Minimum	160.44					
13	Maximum	242.76					
14	Sum	5098.34					
15	Count	25					

6. Return to the worksheet containing the data by clicking on the **8_4_9** sheet tab at the bottom of the screen.

7. At the top of the screen, click **PHStat**. Select **Confidence Intervals → Estimate for the Population Variance**.

8. Complete the Estimate for the Population Variance dialog box with a sample size of **25**, sample standard deviation of **18.32043**, a confidence interval of **99**, and a title of **Tensile Strength**. Click **OK**.

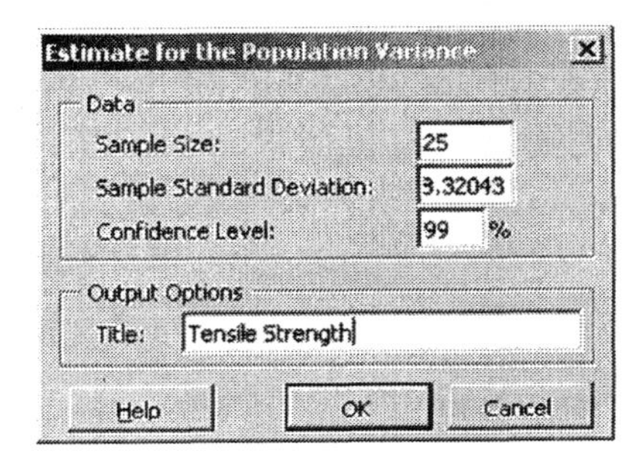

The output is displayed in a worksheet named PVInterval.

	A	B	C	D	E	F	G
1	**Tensile Strength**						
2							
3	**Data**						
4	**Sample Size**	25					
5	**Sample Standard Deviation**	18.32043					
6	**Confidence Level**	99%					
7							
8	Intermediate Calculations						
9	Degrees of Freedom	24					
10	Sum of Squares	8055.316					
11	Single Tail Area	0.005					
12	Lower Chi-Square Value	9.886199					
13	Upper Chi-Square Value	45.55836					
14							
15	**Results**						
16	**Interval Lower Limit for Variance**	176.8131					
17	**Interval Upper Limit for Variance**	814.8042					
18							
19	**Interval Lower Limit for Standard Deviation**	13.29711					
20	**Interval Upper Limit for Standard Deviation**	28.54477					
21							
22	*Assumption:*						
23	**Population from which sample was drawn has an approximate normal distribution.**						

◀

CHAPTER 9

Hypothesis Testing

Section 9.2 Testing a Hypothesis about μ, σ Known

► Exercise 7 (pg. 539) Computing the P-Value

If the PHStat add-in has not been loaded, you will need to load it before continuing. Follow the instructions in Section GS 8.2.

1. Open a new Excel worksheet.

2. At the top of the screen, click **PHStat**. Select **One-Sample Tests → Z Test for the Mean, sigma known**.

3. Complete the Z Test for the Mean dialog box as shown below. Click **OK**.

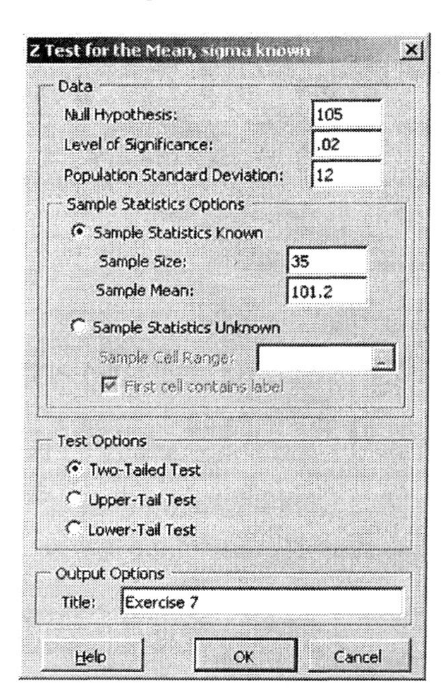

The output is placed in a worksheet named Hypothesis.

	A		B	C	D	E	F	G
1	Exercise 7							
2								
3	Data							
4	Null Hypothesis	μ=	105					
5	Level of Significance		0.02					
6	Population Standard Deviation		12					
7	Sample Size		35					
8	Sample Mean		101.2					
9								
10	Intermediate Calculations							
11	Standard Error of the Mean		2.028370211					
12	Z Test Statistic		-1.873425265					
13								
14	Two-Tailed Test							
15	Lower Critical Value		-2.326341928					
16	Upper Critical Value		2.326341928					
17	*p*-Value		0.061009555					
18	Do not reject the null hypothesis							

► Exercise 19 (pg. 544) Testing the Researcher's Claim, $\alpha = .05$

If the PHStat add-in has not been loaded, you will need to load it before continuing. Follow the instructions in Section GS 8.2.

1. Open a new Excel worksheet.

2. At the top of the screen, click PHStat. Select **One-Sample Tests → Z Test for the Mean, sigma known**.

3. Complete the Z Test for the Mean dialog box as shown below. Click **OK**.

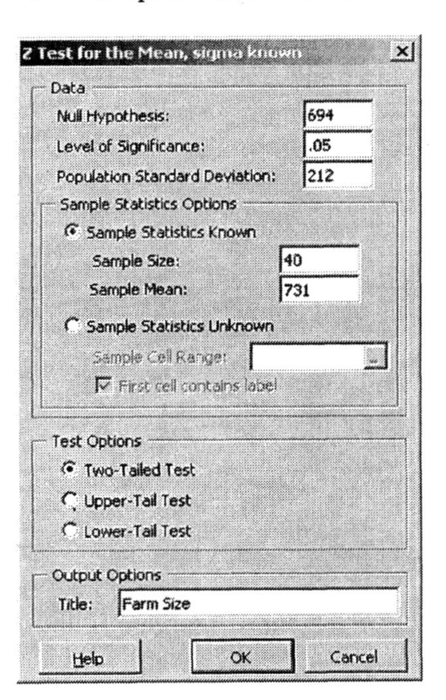

The output is placed in a worksheet named Hypothesis.

	A	B	C	D	E	F	G
1	Farm Size						
2							
3	Data						
4	Null Hypothesis μ=	694					
5	Level of Significance	0.05					
6	Population Standard Deviation	212					
7	Sample Size	40					
8	Sample Mean	731					
9							
10	Intermediate Calculations						
11	Standard Error of the Mean	33.5201432					
12	Z Test Statistic	1.1038139					
13							
14	Two-Tailed Test						
15	Lower Critical Value	-1.959961082					
16	Upper Critical Value	1.959961082					
17	p-Value	0.269673957					
18	Do not reject the null hypothesis						

◀

Section 9.3 Testing a Hypothesis about μ, σ Unknown

▶ Exercise 5 (pg. 551) — Testing the Hypothesis, $\alpha = 0.05$

If the PHStat add-in has not been loaded, you will need to load it before continuing. Follow the instructions in Section GS 8.2.

1. Open a new Excel worksheet.

2. At the top of the screen, click **PHStat**. Select **One-Sample Tests → t Test for the Mean, sigma unknown**.

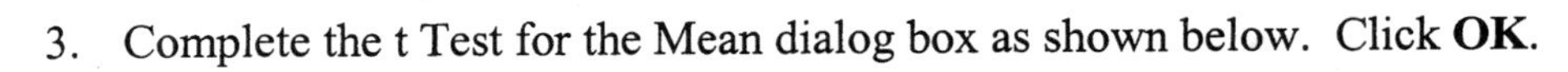

3. Complete the t Test for the Mean dialog box as shown below. Click **OK**.

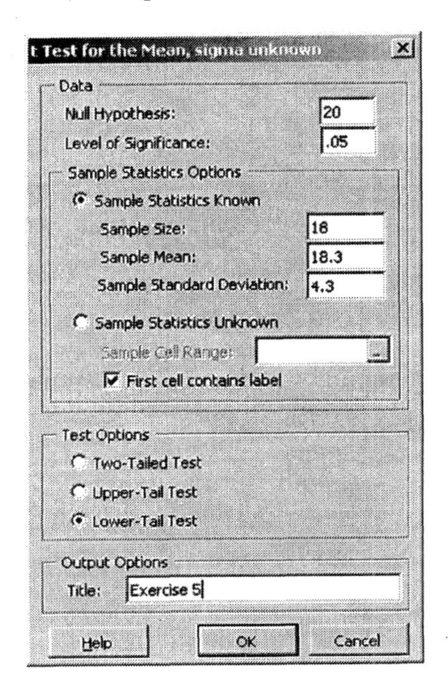

The output is placed in a worksheet named Hypothesis.

	A	B	C	D	E	F	G
1	**Exercise 5**						
2							
3	**Data**						
4	**Null Hypothesis** μ=	**20**					
5	**Level of Significance**	**0.05**					
6	**Sample Size**	**18**					
7	**Sample Mean**	**18.3**					
8	**Sample Standard Deviation**	**4.3**					
9							
10	Intermediate Calculations						
11	Standard Error of the Mean	1.01351972					
12	Degrees of Freedom	17					
13	*t* Test Statistic	-1.677323062					
14							
15	**Lower-Tail Test**						
16	**Lower Critical Value**	**-1.739606432**					
17	***p*-Value**	**0.055884167**					
18	**Do not reject the null hypothesis**						

◀

Section 9.4 Testing a Hypothesis about a Population Proportion

► Exercise 1 (pg. 565)

Testing the Hypothesis p = 0.3

If the PHStat add-in has not been loaded, you will need to load it before continuing. Follow the instructions in Section GS 8.2.

1. Open a new Excel worksheet.

2. At the top of the screen, click **PHStat**. Select **One-Sample Tests → Z Test for the Proportion**.

3. Complete the Z Test for the Proportion dialog box as shown below. Click **OK**.

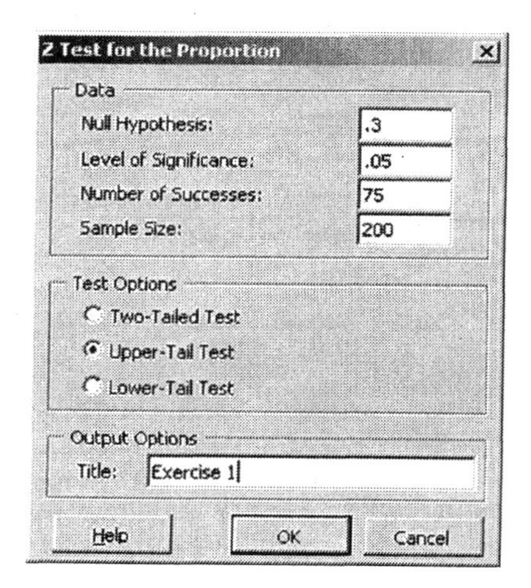

The output is placed in a worksheet named Hypothesis.

	A	B	C	D	E	F	G
1	**Exercise 1**						
2							
3	**Data**						
4	**Null Hypothesis** *p*=	0.3					
5	**Level of Significance**	0.05					
6	**Number of Successes**	75					
7	**Sample Size**	200					
8							
9	Intermediate Calculations						
10	Sample Proportion	0.375					
11	Standard Error	0.032403703					
12	Z Test Statistic	2.314550249					
13							
14	**Upper-Tail Test**						
15	**Upper Critical Value**	1.644853					
16	***p*-Value**	0.010318753					
17	**Reject the null hypothesis**						

◀

Section 9.5 Testing a Hypothesis about σ

▸ Exercise 1 (pg. 573) Testing the Hypothesis $\sigma = 50$

If the PHStat add-in has not been loaded, you will need to load it before continuing. Follow the instructions in Section GS 8.2.

1. Open a new Excel worksheet.

2. At the top of the screen, click **PHStat**. Select **One-Sample Tests → Chi-Square Test for the Variance**.

3. Complete the Chi-Square Test for the Variance dialog box as shown below. Click **OK**.

The Null Hypothesis value of 2500 shown in the dialog box is the variance. The variance was obtained by squaring the numerical value of σ (i.e., 50) that appears in the null hypothesis.

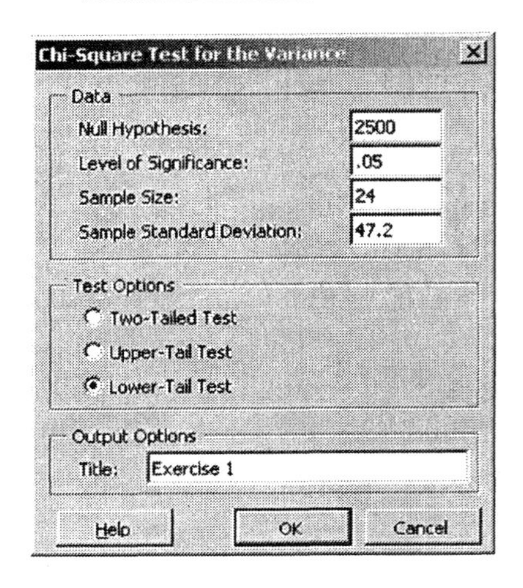

The output is placed in a worksheet named Hypothesis.

	A	B	C	D	E	F	G
1	**Exercise 1**						
2							
3	**Data**						
4	**Null Hypothesis** σ^2=	2500					
5	**Level of Significance**	0.05					
6	**Sample Size**	24					
7	**Sample Standard Deviation**	47.2					
8							
9	Intermediate calculations						
10	Degrees of Freedom	23					
11	Half Area	0.025					
12	Chi-Square Statistic	20.496128					
13							
14	**Lower-Tail Test Results**						
15	Lower Critical Value	13.09050504					
16	***p*-Value**	**0.388171516**					
17	**Do not reject the null hypothesis**						

◀

Inferences on Two Samples

CHAPTER 10

Section 10.1 Inference about Two Means: Dependent Samples

▶ Exercise 9 (pg. 602) — Testing the Difference in the Measurement of Muzzle Velocity

1. Open worksheet "10_1_9" in the Chapter 10 folder. The first few rows are shown below.

	A	B	C	D	E	F	G
1	A	B					
2	793.8	793.2					
3	793.1	793.3					
4	792.4	792.6					

2. At the top of the screen, click **Tools** and select **Data Analysis**.

If Data Analysis does not appear as a choice in the Tools menu, you will need to load the Microsoft Excel Analysis ToolPak add-in. Follow the procedure in Section GS 8.1 before continuing.

3. Select **t-Test: Paired Two Samples** for means. Click **OK**.

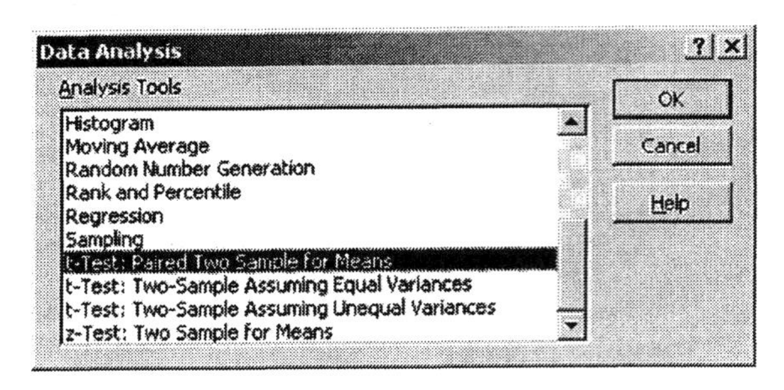

4. Complete the t-Test: Paired Two Sample for Means dialog box as shown below. Click **OK**.

The default value for the Hypothesized Mean Difference is zero. So you can leave this window blank if your hypothesized difference is zero. A checkmark is necessary in the box to the left of Labels if the first row in the variable range is a label rather than a data value.

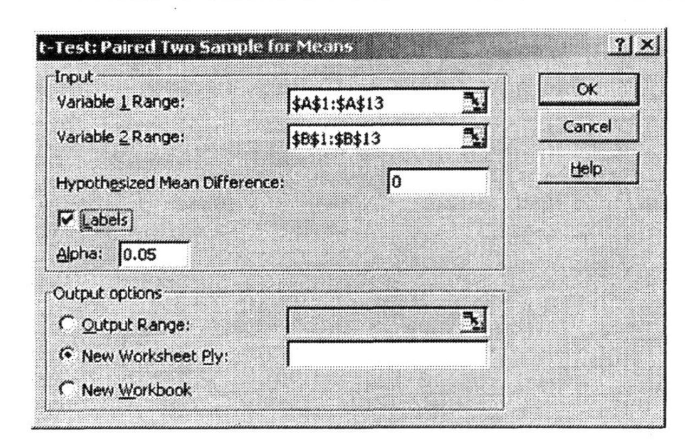

The output is displayed in a new worksheet. Make column A wider so that you can read the labels.

	A	B	C	D	E	F	G
1	t-Test: Paired Two Sample for Means						
2							
3		*A*	*B*				
4	Mean	792.4583	792.3417				
5	Variance	1.979015	2.568106				
6	Observations	12	12				
7	Pearson Correlation	0.958563					
8	Hypothesized Mean Difference	0					
9	df	11					
10	t Stat	0.851726					
11	P(T<=t) one-tail	0.206262					
12	t Critical one-tail	1.795884					
13	P(T<=t) two-tail	0.412525					
14	t Critical two-tail	2.200986					

◀

Section 10.2 Inference about Two Means: Independent Samples

▶ Exercise 9 (pg. 616) Testing a Hypothesis Regarding Two Different Concrete Mix Designs

If the PHStat add-in has not been loaded, you will need to load it before continuing. Follow the instructions in Section GS 8.2.

1. Open worksheet "10_2_9" in the Chapter 10 folder. The first few rows are shown below.

	A	B	C	D	E	F	G
1	Mixture 67	Mixture 67-0-400					
2	3960	4070					
3	3830	4640					
4	2940	5020					

2. At the top of the screen, click **Tools** and select **Data Analysis**.

3. Select **t-Test: Two-Sample Assuming Equal Variances** and click **OK**.

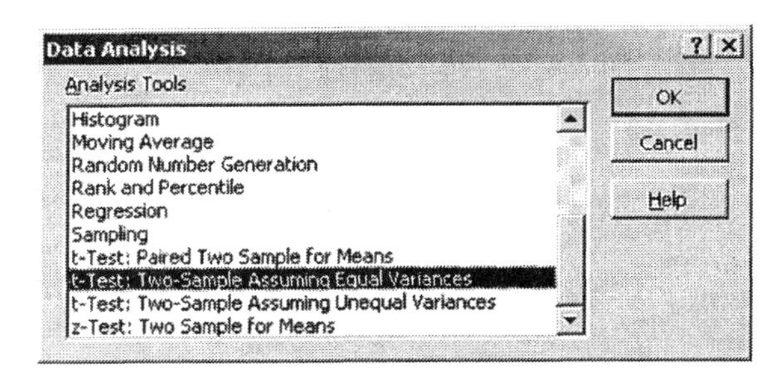

4. Complete the t-Test: Two-Sample Assuming Equal Variances dialog box as shown below. Click **OK**.

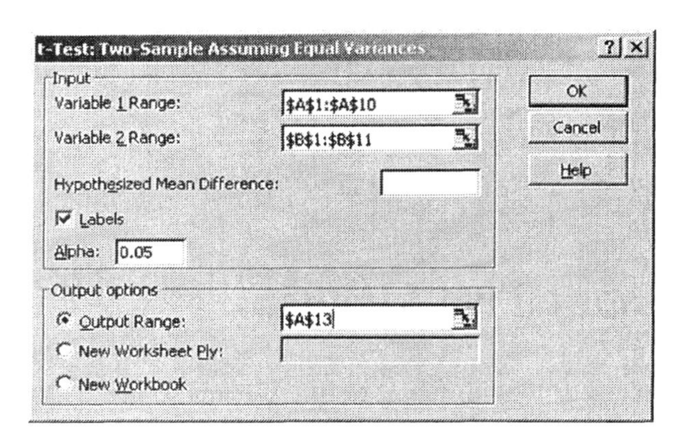

5. The output is placed a couple rows below the data. Make the columns wider so that you can read the entire labels.

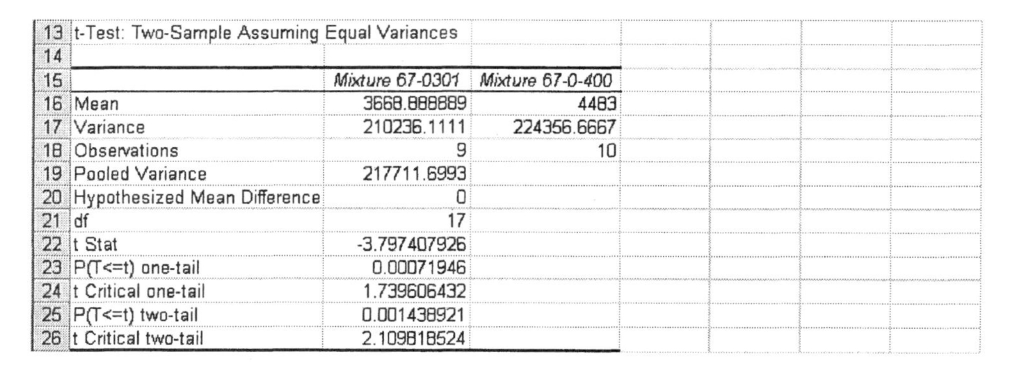

13	t-Test: Two-Sample Assuming Equal Variances		
14			
15		*Mixture 67-0301*	*Mixture 67-0-400*
16	Mean	3668.888889	4483
17	Variance	210236.1111	224356.6667
18	Observations	9	10
19	Pooled Variance	217711.6993	
20	Hypothesized Mean Difference	0	
21	df	17	
22	t Stat	-3.797407926	
23	P(T<=t) one-tail	0.00071946	
24	t Critical one-tail	1.739606432	
25	P(T<=t) two-tail	0.001438921	
26	t Critical two-tail	2.109818524	

6. Next, you will draw box plots of each data set beginning with Mixture 67-0-301. At the top of the screen, click **PHStat**.

7. Select **Descriptive Statistics → Box-and-Whisker Plot**.

8. Complete the Box-and-Whisker Plot dialog box as shown below. Click **OK**.

The Raw Data Cell Range was entered by clicking and dragging over the data range in the worksheet. If you prefer, you can type ***A1:A10***.

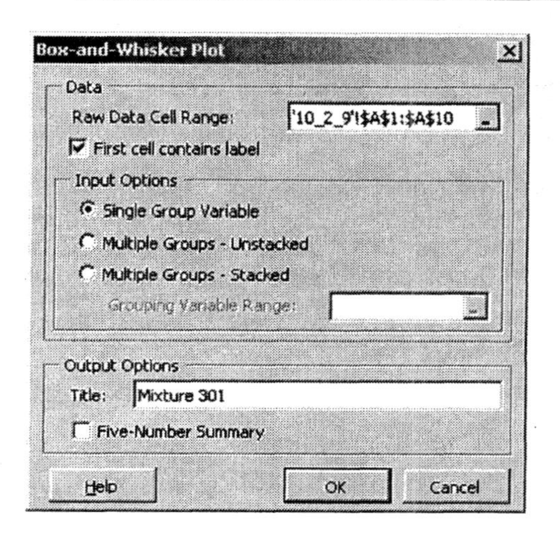

9. The box plot is displayed in a worksheet named BoxWhiskerPlot. Next, you will draw a box plot of the Mixture 67-0-400 data. Return to the worksheet containing the data by clicking on the **10_2_9** sheet tab at the bottom of the screen. At the top of the screen, click **PHStat**.

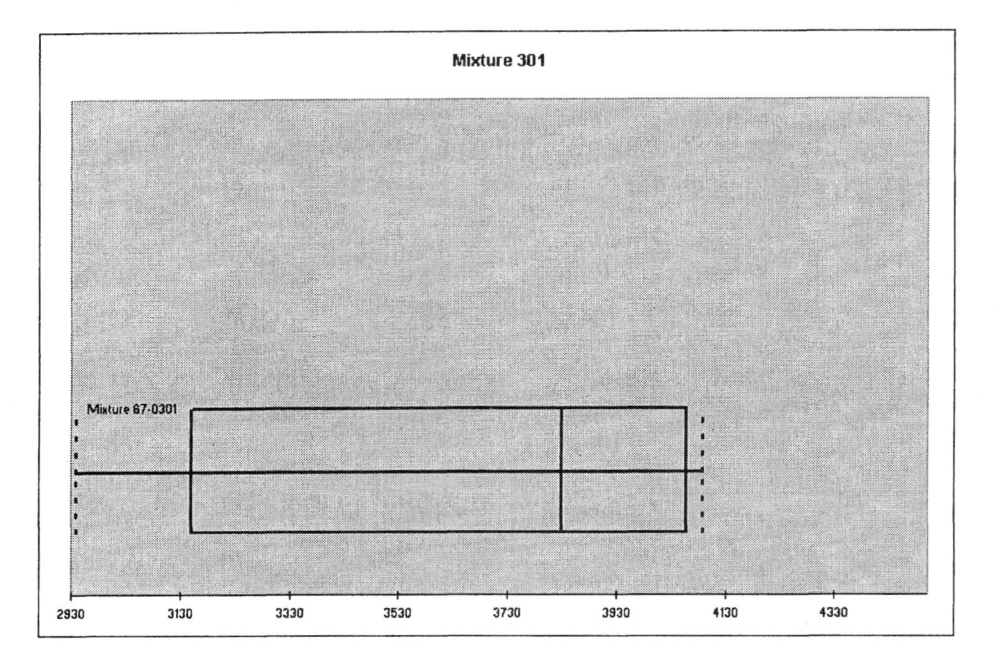

10. Select **Descriptive Statistics → Box-and-Whisker Plot**.

11. Complete the Box-and-Whisker Plot dialog box as shown below. Click **OK**.

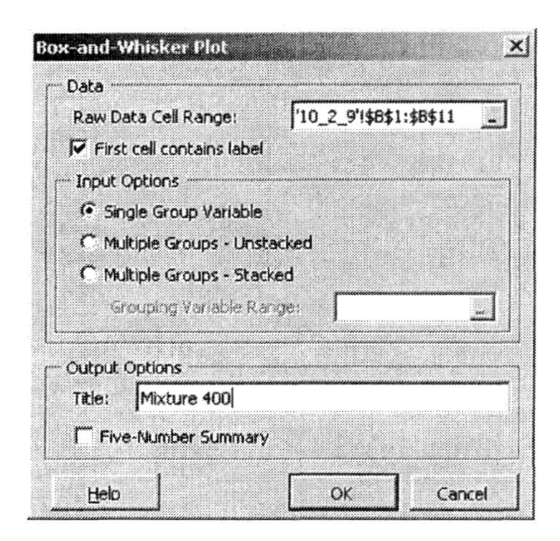

The output is displayed in a worksheet named BoxWhiskerPlot2.

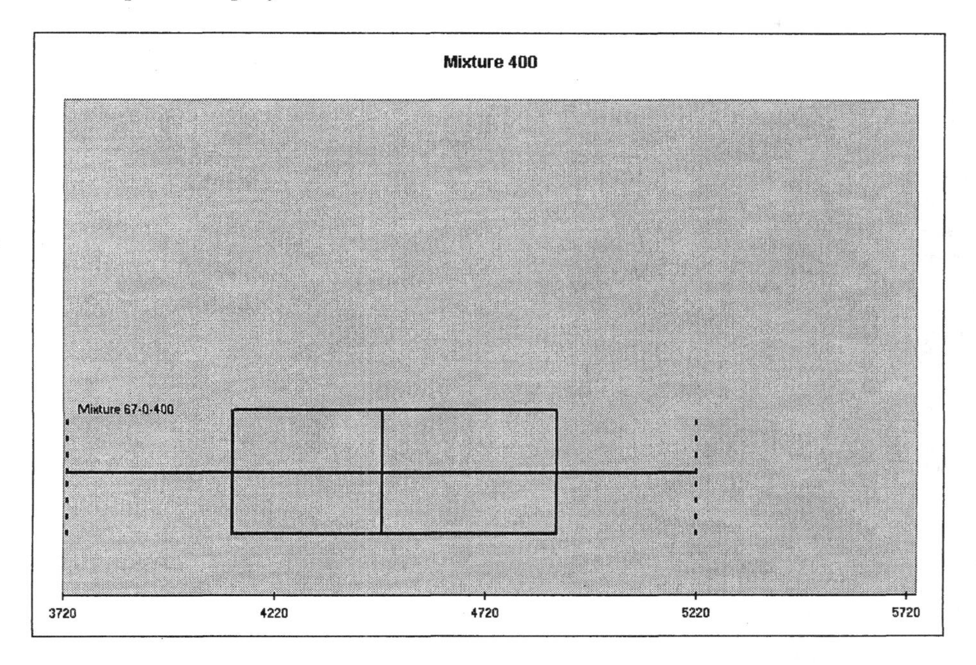

◀

Section 10.3 Inference about Two Population Proportions

► Exercise 11 (pg. 629) | Testing a Hypothesis Regarding Cholesterol

If the PHStat add-in has not been loaded, you will need to load it before continuing. Follow the instructions in Section GS 8.2.

1. Open a new Excel worksheet.

2. At the top of the screen, click **PHStat**.

3. Select **Two-Sample Tests → Z Test for the Difference in Two Proportions**.

4. Complete the Z Test for the Difference in Two Proportions dialog box as shown below. Click **OK**.

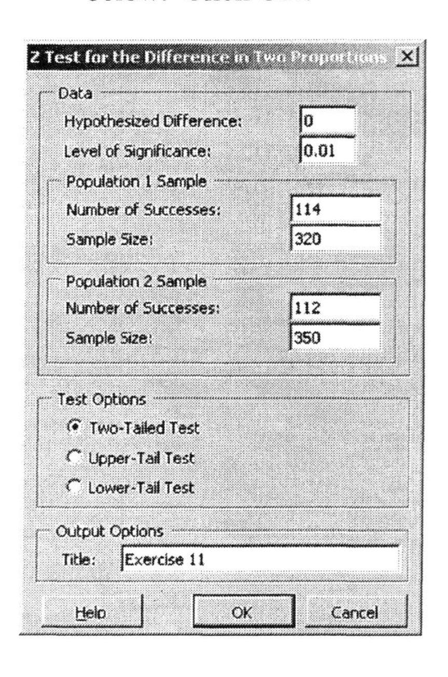

The output is displayed in a worksheet named Hypothesis.

	A	B	C	D	E	F	G
1	Exercise 11						
2							
3	Data						
4	Hypothesized Difference	0					
5	Level of Significance	0.01					
6	Group 1						
7	Number of Successes	114					
8	Sample Size	320					
9	Group 2						
10	Number of Successes	112					
11	Sample Size	350					
12							
13	Intermediate Calculations						
14	Group 1 Proportion	0.35625					
15	Group 2 Proportion	0.32					
16	Difference in Two Proportions	0.03625					
17	Average Proportion	0.337313433					
18	Z Test Statistic	0.991308246					
19							
20	Two-Tailed Test						
21	Lower Critical Value	-2.575834515					
22	Upper Critical Value	2.575834515					
23	*p*-Value	0.321535092					
24	Do not reject the null hypothesis						

◀

Section 10.4 Inference about Two Population Standard Deviations

► Exercise 19 (pg. 642) — Testing a Claim Regarding Wait Time Standard Deviations

1. Open worksheet "10_4_19" in the Chapter 10 folder. The first few rows are shown below.

	A	B	C	D	E	F	G
1	Single Lin	Multiple Lines					
2	1.2	1.1					
3	1.9	3.8					
4	2.1	4.3					

2. At the top of the screen, click **Tools** and select **Data Analysis**.

If Data Analysis does not appear as a choice in the Tools menu, you will need to load the Microsoft Excel Analysis ToolPak add-in. Follow the procedure in Section GS 8.1 before continuing.

---

3. Select **F-Test Two-Sample for Variances** and click **OK**.

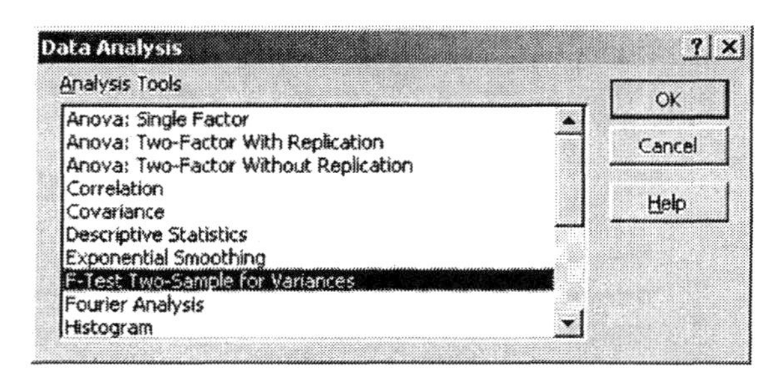

4. Complete the F-Test Two-Sample for Variances dialog box as shown below. Click **OK**.

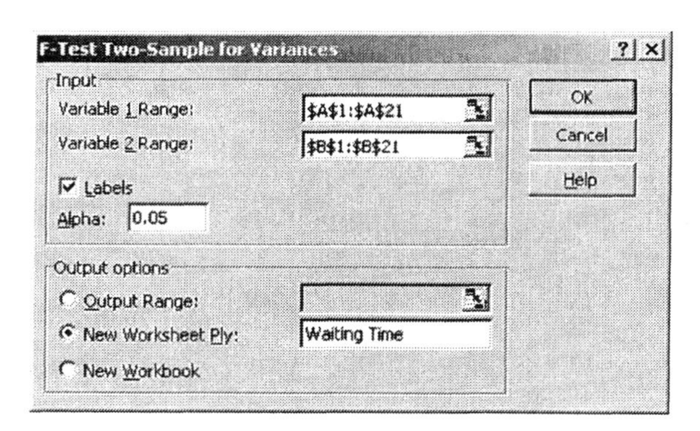

The output is displayed in a worksheet named Waiting Time. Make the columns wider so that you can read the entire labels.

	A	B	C	D	E	F	G
1	F-Test Two-Sample for Variances						
2							
3		*Single Line*	*Multiple Lines*				
4	Mean	2.255	2.475				
5	Variance	0.329973684	1.025131579				
6	Observations	20	20				
7	df	19	19				
8	F	0.321884225					
9	P(F<=f) one-tail	0.008700046					
10	F Critical one-tail	0.461200855					

◀

Chi-Square Procedures

CHAPTER 11

Section 11.2 Contingency Tables; Association

▶ Exercise 1 (pg. 674) — Constructing a Frequency Marginal Distribution and Conditional Distribution

1. Open a new Excel worksheet and enter the data shown below. Begin by calculating the column totals. Click in cell **B4** where you will place the sum of the LT 18 column.

	A	B	C	D	E	F	G
1	Poverty Level	LT 18	18-44	45-64	GE 65	Row Total	
2	Below	8550	4356	1520	958		
3	Above	5884	3741	1756	1943		
4	Column Total						

2. At the top of the screen, click the AutoSum button Σ. You will see =SUM(B2:B3) in cell B4. Press **[Enter]**.

	A	B	C	D	E	F	G
1	Poverty Level	LT 18	18-44	45-64	GE 65	Row Total	
2	Below	8550	4356	1520	958		
3	Above	5884	3741	1756	1943		
4	Column Total	=SUM(B2:B3)					

3. Copy the SUM function in cell B4 to cells C4 through E4. Then click in cell **F2** where you will place the sum of the Below row.

	A	B	C	D	E	F	G
1	Poverty Level	LT 18	18-44	45-64	GE 65	Row Total	
2	Below	8550	4356	1520	958		
3	Above	5884	3741	1756	1943		
4	Column Total	14434	8097	3276	2901		

4. At the top of the screen, click the AutoSum button Σ. You will see =SUM(B2:E2). Press **[Enter]**.

	A	B	C	D	E	F	G
1	Poverty Level	LT 18	18-44	45-64	GE 65	Row Total	
2	Below	8550	4356	1520	958	=SUM(B2:E2)	
3	Above	5884	3741	1756	1943		
4	Column Total	14434	8097	3276	2901		

5. Copy the SUM function in cell F2 to cells F3 through F4. Then click in cell **B5** where you will place the relative frequency for the LT 18 column.

	A	B	C	D	E	F	G
1	Poverty Level	LT 18	18-44	45-64	GE 65	Row Total	
2	Below	8550	4356	1520	958	15384	
3	Above	5884	3741	1756	1943	13324	
4	Column Total	14434	8097	3276	2901	28708	
5							

6. Key in the formula shown below to calculate the relative marginal frequency. Note that the dollar signs are necessary here. You want to make the F4 reference absolute so that it will not change when it is copied. Press **[Enter]**.

	A	B	C	D	E	F	G
1	Poverty Level	LT 18	18-44	45-64	GE 65	Row Total	
2	Below	8550	4356	1520	958	15384	
3	Above	5884	3741	1756	1943	13324	
4	Column Total	14434	8097	3276	2901	28708	
5		=B4/$F$4					

7. Copy the formula in cell B5 to cells C5 through E5. Then click in cell **G2** where you will place the relative frequency for the Below row.

	A	B	C	D	E	F	G
1	Poverty Level	LT 18	18-44	45-64	GE 65	Row Total	
2	Below	8550	4356	1520	958	15384	
3	Above	5884	3741	1756	1943	13324	
4	Column Total	14434	8097	3276	2901	28708	
5		0.502787	0.282047	0.114115	0.101052		

8. Key in the formula shown below to calculate the relative marginal frequency. Press **[Enter]**.

	A	B	C	D	E	F	G
1	Poverty Level	LT 18	18-44	45-64	GE 65	Row Total	
2	Below	8550	4356	1520	958	15384	=F2/$F$4
3	Above	5884	3741	1756	1943	13324	
4	Column Total	14434	8097	3276	2901	28708	
5		0.502787	0.282047	0.114115	0.101052		

9. Copy the formula in cell G2 to cell G3.

	A	B	C	D	E	F	G
1	Poverty Level	LT 18	18-44	45-64	GE 65	Row Total	
2	Below	8550	4356	1520	958	15384	0.535879
3	Above	5884	3741	1756	1943	13324	0.464121
4	Column Total	14434	8097	3276	2901	28708	
5		0.502787	0.282047	0.114115	0.101052		

10. You now will construct a conditional distribution by poverty level. Copy cells **A1:E3** to an area two rows below the previously constructed table.

	A	B	C	D	E	F	G
1	Poverty Level	LT 18	18-44	45-64	GE 65	Row Total	
2	Below	8550	4356	1520	958	15384	0.535879
3	Above	5884	3741	1756	1943	13324	0.464121
4	Column Total	14434	8097	3276	2901	28708	
5		0.502787	0.282047	0.114115	0.101052		
6							
7							
8	Poverty Level	LT 18	18-44	45-64	GE 65		
9	Below	8550	4356	1520	958		
10	Above	5884	3741	1756	1943		

11. Begin with the conditional probabilities in the Below row. Click in cell **B9** where you will place the LT 18 probability. Key in the formula shown below. Press **[Enter]**.

8	Poverty Level	LT 18	18-44	45-64	GE 65		
9	Below	=B2/$F$2	4356	1520	958		
10	Above	5884	3741	1756	1943		

12. Copy the formula in cell B9 to cells C9 through E9. Then click in cell **B10** where you will enter a formula to calculate the conditional probability for LT 18 in the Above row.

8	Poverty Level	LT 18	18-44	45-64	GE 65		
9	Below	0.555772	0.283151	0.098804	0.062272		
10	Above	5884	3741	1756	1943		

13. Key in the formula shown below. Press **[Enter]**.

8	Poverty Level	LT 18	18-44	45-64	GE 65		
9	Below	0.555772	0.283151	0.098804	0.062272		
10	Above	=B3/$F$3	3741	1756	1943		

14. Copy the formula in cell B10 to cells C10 through E10. The completed conditional probability distribution is shown below.

8	Poverty Level	LT 18	18-44	45-64	GE 65		
9	Below	0.555772	0.283151	0.098804	0.062272		
10	Above	0.441609	0.280772	0.131792	0.145827		

15. You will now draw a bar graph of this conditional distribution. At the top of the screen, click **Insert** and select **Chart**.

16. Under Chart type, select **Column**. Under Chart sub-type, select the leftmost diagram in the top row. Click **Next>**.

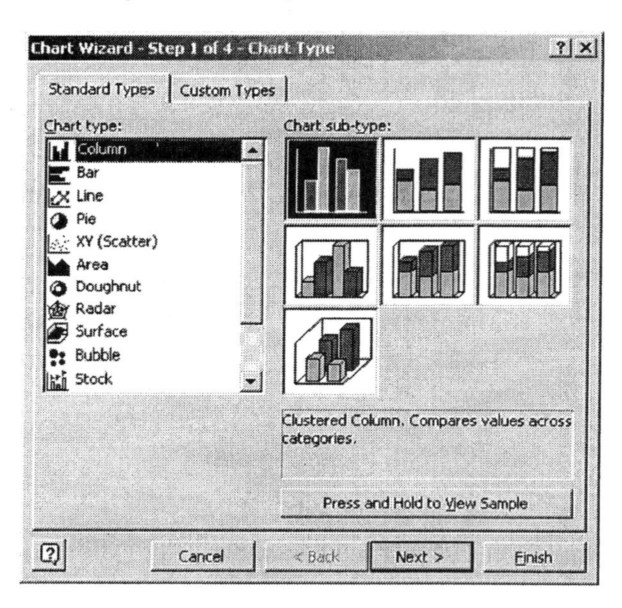

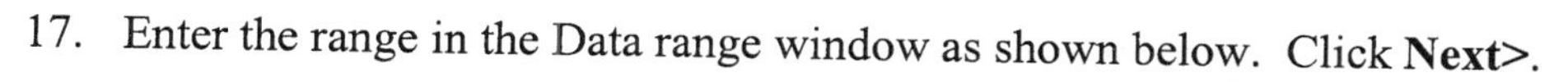

17. Enter the range in the Data range window as shown below. Click **Next>**.

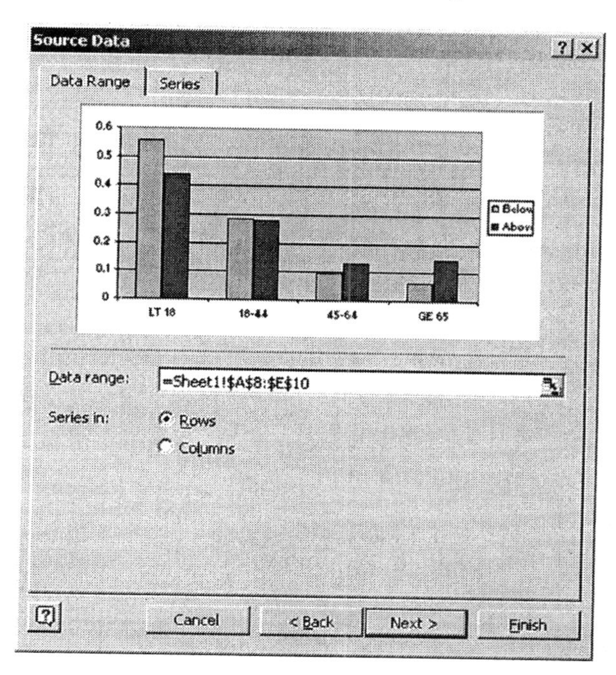

18. Click the **Titles** tab at the top of the Chart Options dialog box. For the Chart title, enter **Conditional Distribution by Poverty Level**. For the Category (X) axis, enter **Age**. For the Value (Y) axis, enter **Relative Frequency**. Click **Next>**.

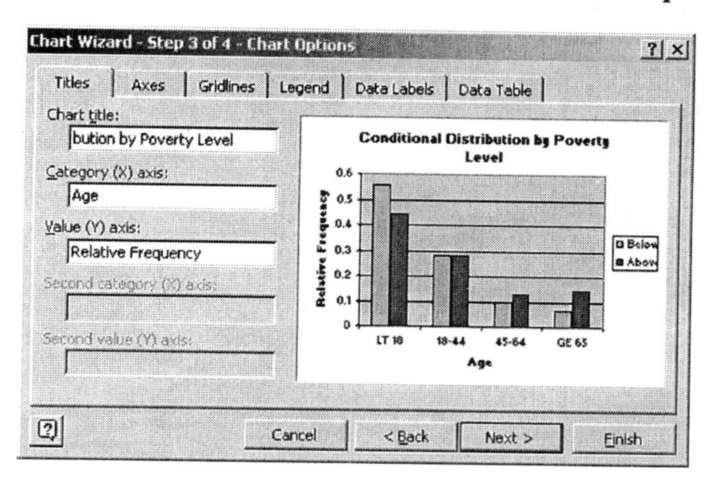

19. The Chart Location dialog box presents two options for placement of the chart. Select **As new sheet**. Click **Finish**.

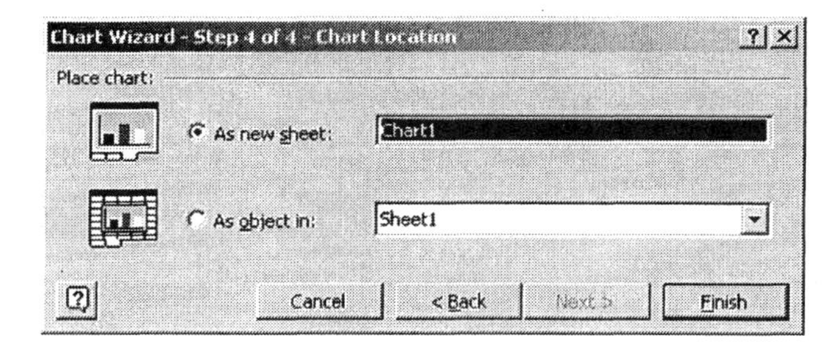

The completed chart is shown below.

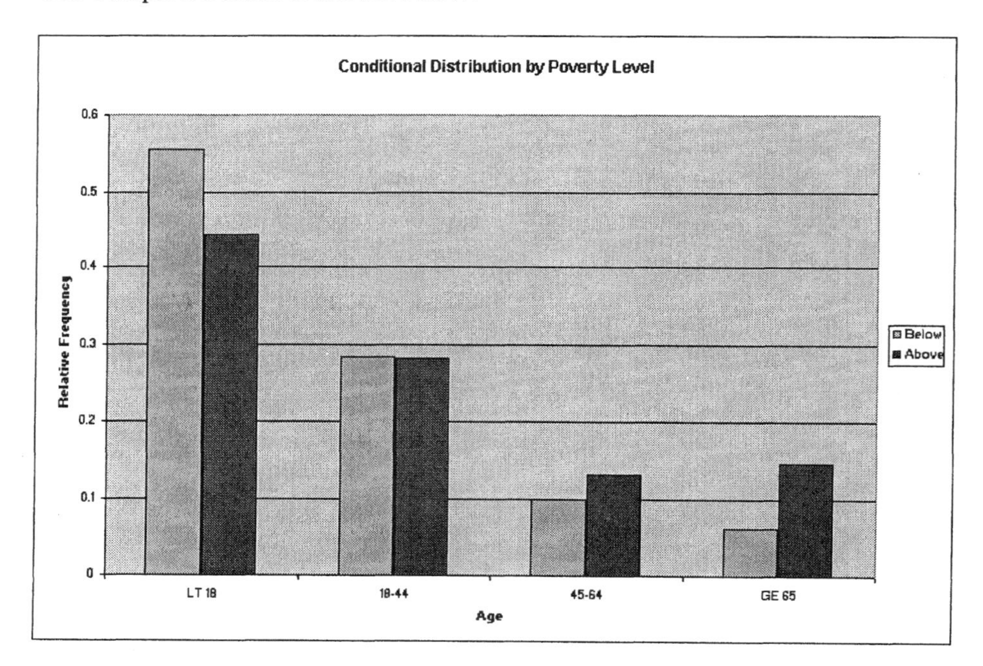

Section 11.3 Chi-Square Test for Independence; Homogeneity of Proportions

► Exercise 1 (pg. 687) | Computing the Value of the Chi-Square Test

If the PHStat add-in has not been loaded, you will need to load it before continuing. Follow the instructions in Section GS 8.2.

1. Open a new Excel worksheet.

2. At the top of the screen, click **PHStat**. Select **Multiple-Sample Tests → Chi-Square Test.**

3. Complete the Chi-Square Test dialog box as shown below. Click **OK**.

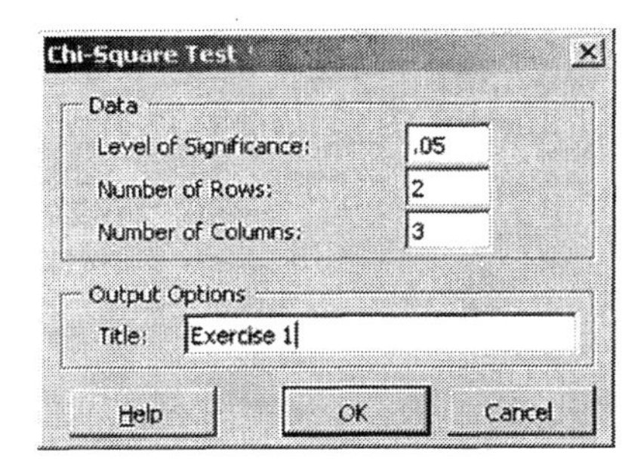

4. Enter observed frequencies in the top section of the worksheet as shown below.

	A	B	C	D	E
1	Exercise 1				
2					
3	Observed Frequencies				
4		Column variable			
5	Row variable	C1	C2	C3	Total
6	R1	34	43	52	129
7	R2	18	21	17	56
8	Total	52	64	69	185

5. Scroll down to see the results of the chi-square test.

23	Results	
24	Critical Value	5.991476
25	Chi-Square Test Statistic	1.698215
26	*p*-Value	0.427797
27	Do not reject the null hypothesis	

◄

Inference on the Least-Squares Regression Model; ANOVA

CHAPTER 12

Section 12.1 Inference about the Least-Squares Regression Model

▸ Example 1 (pg. 706) — Drawing a Scatter Diagram and Finding the Least-Squares Regression Equation

1. Open worksheet "12_1_Ex1" in the Chapter 12 folder. The first few lines are shown below.

If you would like to draw a scatter diagram of these data, see Section 4.1 of this manual for instructions.

	A	B	C	D	E	F	G
1	AGE	TOTAL CHOLESTEROL					
2	25	180					
3	25	195					
4	28	186					

2. At the top of the screen, click **Tools** and select **Data Analysis**.

If Data Analysis does not appear as a choice in the Tools menu, you will need to load the Microsoft Excel Analysis ToolPak add-in. Follow the procedure in Section GS 8.1 before continuing.

3. In the Data Analysis dialog box, select **Regression** and click **OK**.

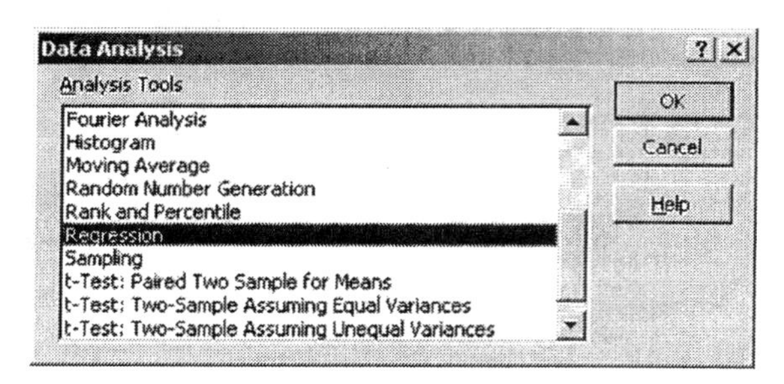

4. Complete the Regression dialog box as shown below. Click **OK**.

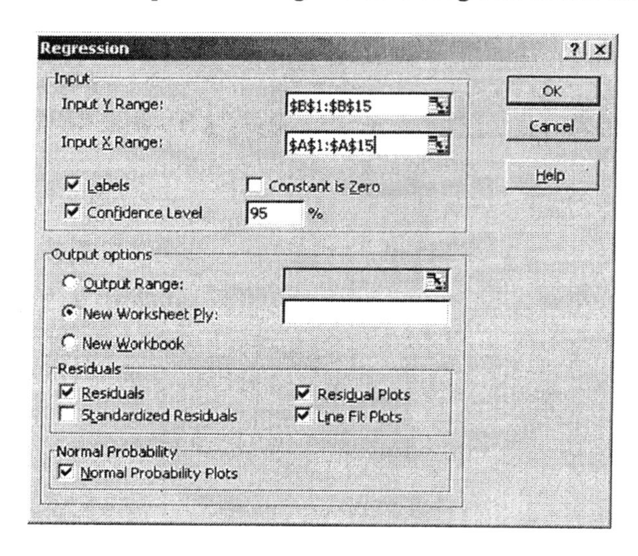

Because you requested many output options, you will need to scroll down and to the right to view all the output. The correlation (Multiple R) is 0.7178. The coefficient of determination (R Square) is 0.5153. The least-squares regression equation is $\hat{y} = 151.3537 + 1.3991x$.

	A	B	C	D	E	F	G
1	SUMMARY OUTPUT						
2							
3	*Regression Statistics*						
4	Multiple R	0.717811					
5	R Square	0.515252					
6	Adjusted R	0.474856					
7	Standard E	19.48054					
8	Observatio	14					
9							
10	ANOVA						
11		*df*	*SS*	*MS*	*F*	*ignificance F*	
12	Regressior	1	4840.462	4840.462	12.75514	0.003842	
13	Residual	12	4553.895	379.4912			
14	Total	13	9394.357				
15							
16		*Coefficients*	*tandard Err*	*t Stat*	*P-value*	*Lower 95%*	*Upper 95%*
17	Intercept	151.3537	17.28376	8.756987	1.47E-06	113.6956	189.0117
18	AGE	1.399064	0.391737	3.571433	0.003842	0.545542	2.252587

◄

▶ Exercise 11 (pg. 720)

Testing the Claim That a Linear Relation Exists

1. Open worksheet "12_1_11" in the Chapter 12 folder. The first few lines are shown below.

	A	B	C	D	E	F	G
1	Month	Rate of Re	Rate of Return in Cisco Systems				
2	Sep-99	-2.9	1.1				
3	Oct-99	6.3	7.9				
4	Nov-99	1.9	2.1				

2. At the top of the screen, click **Tools** and select **Data Analysis.**

If Data Analysis does not appear as a choice in the Tools menu, you will need to load the Microsoft Excel Analysis ToolPak add-in. Follow the procedure in Section GS 8.1 before continuing.

3. In the Data Analysis dialog box, select **Regression** and click **OK**.

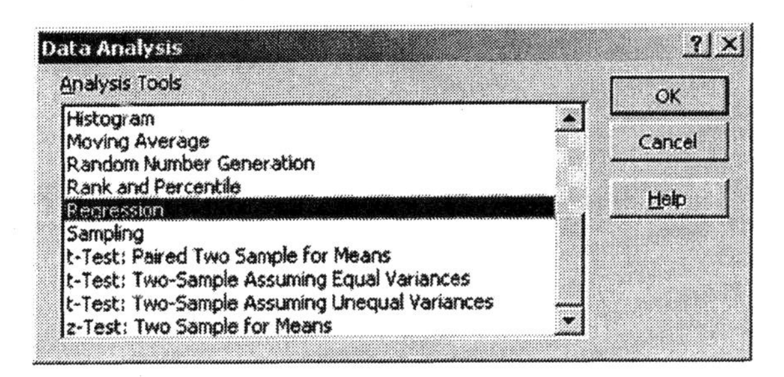

4. Complete the Regression dialog box as shown below. Click **OK**.

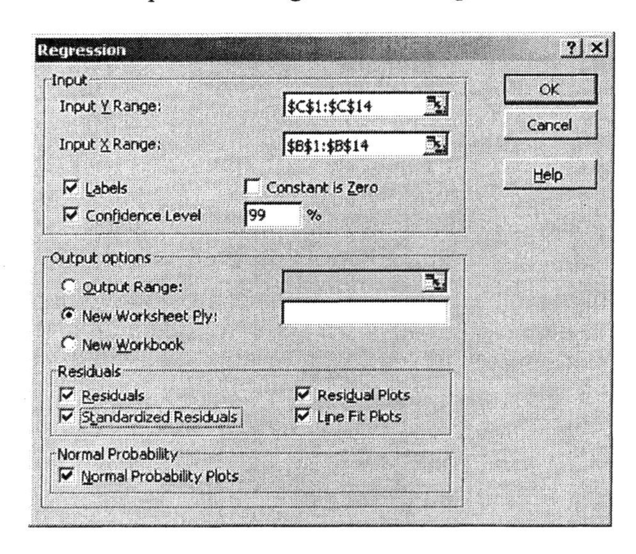

Because you requested many output options, you will need to scroll down and to the right to view all the output. The intercept is 2.0336. The slope is 1.6363. The standard error of the slope is 0.6265. The lower bound of the 99% confidence interval about the slope is –0.3093. The upper bound is 3.5820.

16		Coefficient	Standard Err	t Stat	P-value	Lower 95%	Upper 95%	Lower 99.0%	Upper 99.0%
17	Intercept	2.033584	2.99691	0.67856	0.51144	-4.56257	8.629741	-7.27426	11.34143
18	Rate of Re	1.636341	0.626451	2.612081	0.024168	0.257531	3.015151	-0.3093	3.581983

The correlation (Multiple R) is 0.6187. The coefficient of determination (R Square) is 0.3828. The standard error is 10.6649.

3	Regression Statistics	
4	Multiple R	0.618723
5	R Square	0.382819
6	Adjusted R	0.326711
7	Standard E	10.66491
8	Observatio	13

◀

Section 12.2 Confidence and Prediction Intervals

► Exercise 1 (pg. 727)

Constructing a 95% Confidence Interval About the Mean Value of *y*

If the PHStat add-in has not been loaded, you will need to load it before continuing. Follow the instructions in in Section GS 8.2.

1. Open a new Excel worksheet and enter the X and Y values shown below.

	A	B	C	D	E	F	G
1	X	Y					
2	3	4					
3	4	6					
4	5	7					
5	7	12					
6	8	14					

2. At the top of the screen, click **PHStat**.

3. Select **Regression → Simple Linear Regression**.

4. Complete the Simple Linear Regression dialog box as shown below. Click **OK**.

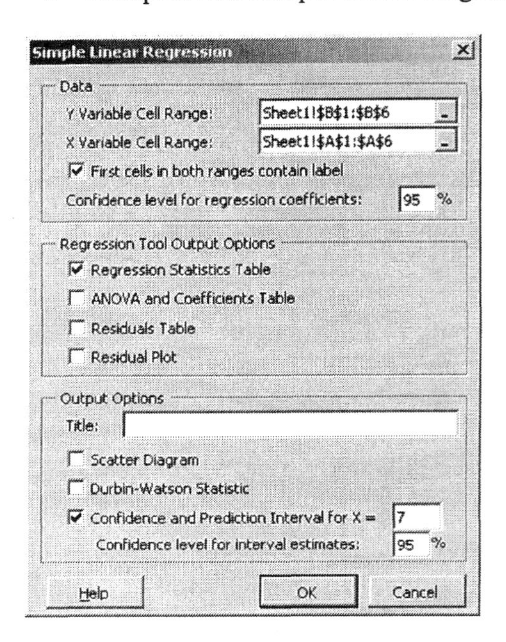

The predicted mean value of y is 11.8372. The 95% confidence interval about the mean value is 10.8722 to 12.8022.

15	Average Predicted Y (YHat)	11.83721
16		
17	**For Average Predicted Y (YHat)**	
18	Interval Half Width	0.964978
19	**Confidence Interval Lower Limit**	**10.87223**
20	**Confidence Interval Upper Limit**	**12.80219**

The 95% prediction interval about the value of y for $x = 7$ is 9.9397 to 13.7347.

22	**For Individual Response Y**	
23	Interval Half Width	1.897518
24	**Prediction Interval Lower Limit**	**9.939691**
25	**Prediction Interval Upper Limit**	**13.73473**

◀

Section 12.3 One-way Analysis of Variance

► Exercise 1 (pg. 738) Testing Whether the Mean Number of Plants for Each Plot Are Equal

1. Open worksheet "3_1_5" in the Chapter 3 folder. The first few lines are shown below.

	A	B	C	D	E	F	G
1	Sludge Plo	Spring Dis	No Till	Spring Chi	Great Lakes BT		
2	25	32	30	30	28		
3	27	30	26	32	32		
4	33	33	29	26	27		

2. At the top of the screen, click **Tools** and select **Data Analysis**.

If Data Analysis does not appear as a choice in the Tools menu, you will need to load the Microsoft Excel Analysis ToolPak add-in. Follow the procedure in Section GS 8.1 before continuing.

3. Select **Anova: Single Factor** and click **OK**.

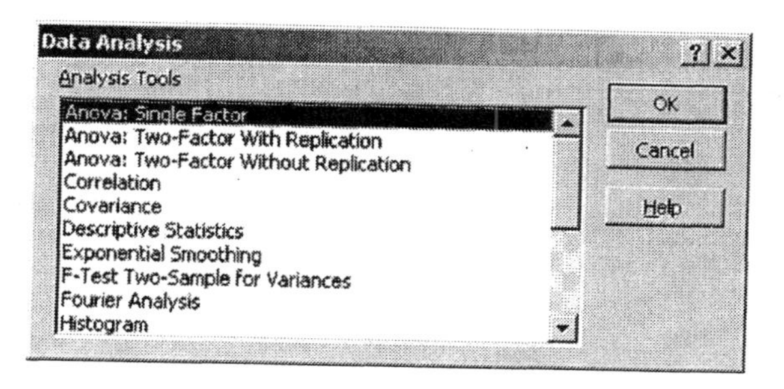

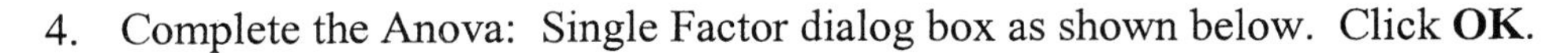

4. Complete the Anova: Single Factor dialog box as shown below. Click **OK**.

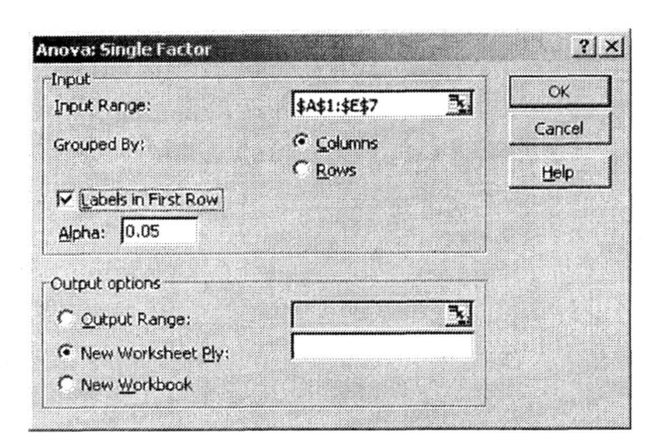

The output is displayed in a new worksheet.

	A	B	C	D	E	F	G
1	Anova: Single Factor						
2							
3	SUMMARY						
4	*Groups*	*Count*	*Sum*	*Average*	*Variance*		
5	Sludge Plc	6	170	28.33333	7.866667		
6	Spring Dis	6	198	33	3.2		
7	No Till	6	171	28.5	6.7		
8	Spring Chi	6	176	29.33333	4.666667		
9	Great Lake	6	173	28.83333	3.766667		
10							
11							
12	ANOVA						
13	*rce of Varia*	*SS*	*df*	*MS*	*F*	*P-value*	*F crit*
14	Between G	90.2	4	22.55	4.303435	0.008734	2.758711
15	Within Gro	131	25	5.24			
16							
17	Total	221.2	29				

◀

▶ Exercise 7 (pg. 743) Testing Whether the Chest Compression Means are Equal

1. Open worksheet "12_3_7" in the Chapter 12 folder.

2. When using Excel to carry out a one-way analysis of variance, the columns of data must be right next to each other. Copy the Chest Compression data to an area in the worksheet that is two rows below the complete data set. Also copy the labels to row 11 as shown below.

	A	B	C	D	E	F	G
1	Large Fam	Chest Con	Passenger	Chest Con	Midsize Ut	Compression	
2	Chevrolet L	34	Toyota Sie	37	Mercedes	42	
3	Ford Tauru	28	Honda Ody	29	Toyota 4R	28	
4	Buick LeS	28	Ford Wind	27	Mitsubishi	39	
5	Chevrolet I	26	Mazda MF	30	Nissan Xte	40	
6	Chrysler LI	22	Chevrolet A	34	Ford Explc	29	
7	Pontiac Gr	34	Nissan Qu	39	Jeep Gran	28	
8	Dodge Intr	24	Pontiac Tr	23	Nissan Pa	36	
9							
10							
11	Large Fam	Passenger	Midsize Utility Vehicles				
12	34	37	42				
13	28	29	28				
14	28	27	39				
15	26	30	40				
16	22	34	29				
17	34	39	28				
18	24	23	36				

3. At the top of the screen, click **Tools** and select **Data Analysis**.

If Data Analysis does not appear as a choice in the Tools menu, you will need to load the Microsoft Excel Analysis ToolPak add-in. Follow the procedure in Section GS 8.1 before continuing.

---

4. Select **Anova: Single Factor** and click **OK**.

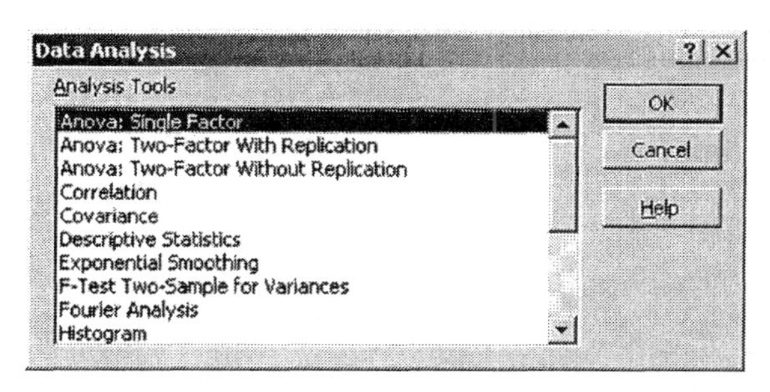

5. Complete the Anova: Single Factor dialog box as shown below. Click **OK**.

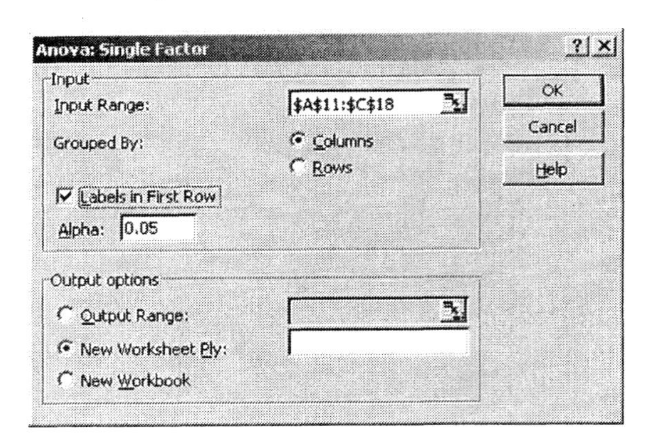

The output is displayed in a new worksheet.

	A	B	C	D	E	F	G
1	Anova: Single Factor						
2							
3	SUMMARY						
4	*Groups*	*Count*	*Sum*	*Average*	*Variance*		
5	Large Fam	7	196	28	21.33333		
6	Passenger	7	219	31.28571	32.2381		
7	Midsize Ut	7	242	34.57143	37.28571		
8							
9							
10	ANOVA						
11	*rce of Varia*	*SS*	*df*	*MS*	*F*	*P-value*	*F crit*
12	Between G	151.1429	2	75.57143	2.495283	0.110536	3.554561
13	Within Gro	545.1429	18	30.28571			
14							
15	Total	696.2857	20				

◀

Nonparametric Statistics

CHAPTER 13

Section 13.7 Kruskal-Wallis Test of One-Way Analysis of Variance

► Exercise 7 (pg. 812)

Testing the Claim That the Distribution for Each Stimulus Is the Same, $\alpha = 0.01$

If the PHStat add-in has not been loaded, you will need to load it before continuing. Follow the instructions in Section GS 8.2. ***Note that PHStat's Kruskal-Wallis procedure can handle a maximum of four groups. If your problem has more than four groups, the Kruskal-Wallis output will not be accurate.***

1. Open worksheet "13_7_7" in the Chapter 13 folder. The first few lines are shown below.

	A	B	C	D	E	F	G
1	Simple	Go/No Go	Choice				
2	0.43	0.588	0.561				
3	0.498	0.375	0.498				
4	0.48	0.409	0.519				

2. At the top of the screen, click **PHStat**. Select **Multiple-Sample Tests → Kruskal-Wallis Rank Test**.

3. Complete the Kruskal-Wallis Rank Test dialog box as shown below. Click **OK**.

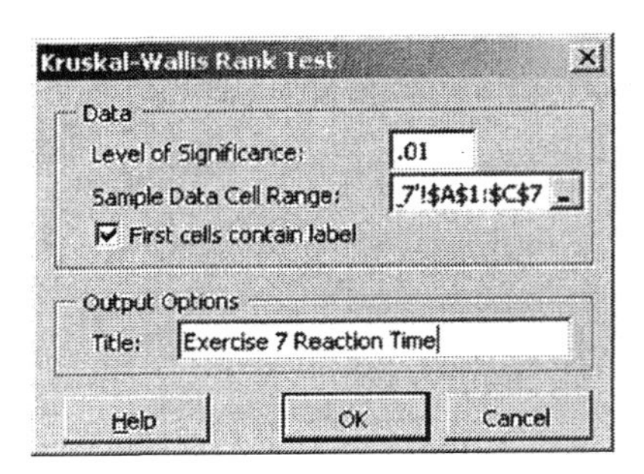

The output is displayed in a worksheet named Kruskal.

	A	B	C	D	E	F	G
1	**Exercise 7 Reaction Time**						
2							
3	**Data**						
4	**Level of Significance**	**0.01**					
5							
6	Group 1						
7	Sum of Ranks	37.5					
8	Sample Size	6					
9	Group 2						
10	Sum of Ranks	54					
11	Sample Size	6					
12	Group 3						
13	Sum of Ranks	79.5					
14	Sample Size	6					
15							
16	Intermediate Calculations						
17	Sum of Squared Ranks/Sample Size	1773.75					
18	Sum of Sample Sizes	18					
19	Number of groups	3					
20	*H* Test Statistic	5.236842					
21							
22	**Test Result**						
23	**Critical Value**	**9.210351**					
24	***p*-Value**	**0.072918**					
25	**Do not reject the null hypothesis**						

◀